ROAD TRANSPORT RESEARCH

roadside noise abatement

REPORT PREPARED BY
AN OECD SCIENTIFIC EXPERT GROUP

ORGANISATION FOR ECONOMIC CO-OPERATION AND DEVELOPMENT

ORGANISATION FOR ECONOMIC CO-OPERATION AND DEVELOPMENT

Pursuant to Article 1 of the Convention signed in Paris on 14th December 1960, and which came into force on 30th September 1961, the Organisation for Economic Co-operation and Development (OECD) shall promote policies designed:

— to achieve the highest sustainable economic growth and employment and a rising standard of living in Member countries, while maintaining financial stability, and thus to contribute to the development of the world economy;
— to contribute to sound economic expansion in Member as well as non-member countries in the process of economic development; and
— to contribute to the expansion of world trade on a multilateral, non-discriminatory basis in accordance with international obligations.

The original Member countries of the OECD are Austria, Belgium, Canada, Denmark, France, Germany, Greece, Iceland, Ireland, Italy, Luxembourg, the Netherlands, Norway, Portugal, Spain, Sweden, Switzerland, Turkey, the United Kingdom and the United States. The following countries became Members subsequently through accession at the dates indicated hereafter: Japan (28th April 1964), Finland (28th January 1969), Australia (7th June 1971), New Zealand (29th May 1973) and Mexico (18th May 1994). The Commission of the European Communities takes part in the work of the OECD (Article 13 of the OECD Convention).

Publié en français sous le titre :
LA RÉDUCTION DU BRUIT AUX ABORDS DES VOIES ROUTIÈRES

FOREWORD

The Programme centres on road and road transport research, while taking into account the impacts of intermodal aspects on the road transport system as a whole. It is geared towards a technico-economic approach to solving key road transport issues identified by Member countries. The Programme has two main fields of activity:

- International research and policy assessments of road and road transport issues to provide scientific support for decisions by Member governments and international governmental organisations;
- Technology transfer and information exchange through two databases - the International Road Research Documentation (IRRD) scheme and the International Road Traffic and Accident Database (IRTAD).

Its mission is to:

- enhance innovative research through international co-operation and networking;
- undertake joint policy analyses and prepare technology reviews of critical road transport issues;
- promote the exchange of scientific and technical information in the transport sector and contribute to road technology transfer in OECD Member and non-member countries.

The scientific and technical activities concern:

- Infrastructure research;
- Road traffic and intermodal transport;
- Environment/transport interactions;
- Traffic safety research;
- Strategic research planning.

ABSTRACT

IRRD No. 870922

The study carried out by an OECD Group of experts reviews the current state-of-the-art and national experience with noise abatement techniques for new and existing roads. The first two Chapters describe problem areas, current practice and noise limits including the present status of national standards. Chapter III on "Assessment and Measurement" focuses on measurement methods and mathematical models as they apply to new and existing infrastructure. Chapter IV on "Anti-noise Design and Layout" deals with road engineering structures (bridges, embankments, tunnels), designing new structures and modifying existing ones. "Low Noise Road Surfacings" are discussed in Chapter V assessing their effectiveness, reliability and serviceability from the safety and environment point of view. Chapter VI on "Noise Barriers" evaluates the various types of barriers used in OECD Member countries considering problems of aesthetics, environment and landscaping. Chapter VII on "Integration of Measures and Costs" discusses practical implementation questions and a methodology for evaluating the costs of simple and/or combined noise abatement techniques. The Report provides a multidisciplinary approach useful for road professionals, environmental planners and transport managers in readjusting road environment policies and programmes.

Field classification: Environment

Field codes: 15

Key Words: Contact (tyre/road); cost; forecast; highway; legislation; mathematical model; measurement; noise barrier; prevention; sound; sound level; surfacing; traffic; traffic control.

TABLE OF CONTENTS

EXECUTIVE SUMMARY

THE PROBLEM OF ROAD NOISE

The assessment and control of environmental impacts of road infrastructures is a relatively recent task for road designers, constructors and managers. The variety of impacts and factors to be considered has been described more generally in a previous 1994 OECD review entitled "Environmental Impact Assessment of Roads". The present Report addresses the most well-known and widespread nuisance, namely roadside noise. It directly concerns the quality of life, mainly in densely populated areas, in which high volumes of road traffic prevail. Its origins and propagation depend on the interaction between three factors:

- vehicles: type, number and speed;
- road structure: its design, construction and materials;
- the environment neighbouring the road-structure, its components and receptors i.e. the characteristics of buildings and the number of inhabitants.

Road noise is a complex phenomenon mainly because of its human sensorial effects and thus has a non-deterministic component, but also due to the fact that it is physically difficult to measure. Its intensity varies with the distance between the source and the receptor and with the environmental conditions influencing noise propagation. Notwithstanding this difficulty, however, its environmental impact is the least difficult to evaluate, given the considerable amount of practical experience that has been gathered.

At international level, regulations and standards exist, mainly for the acoustic properties of materials and vehicles and their measurement. However, the overall management and control of the noise phenomenon, especially in the case of roads, are less well defined; there are numerous and different approaches both in regard to methods of evaluation and regulations.

THE STANDARDS TO ADOPT

One of the specific factors in road noise control is the responsibility issue. In fact the persons responsible for roadside noise are not the owners/authorities of the infrastructure, but the motorists themselves. This means that no immediate action can be taken to determine who is responsible for reducing the noise level in the absence of specific legislation. In the case of new roads, road builders are always required to ensure that the levels of noise are reduced to a minimum. However, as regards existing roads, the road authority is not always required to reduce the levels of acoustic pollution.

It should also be added that in the case of existing roads, even the onus of the acoustic pollution often resides with the contractor having built the "receptors" (buildings, houses), e.g:

- in areas where major road infrastructures were already present;
- housing with inadequate acoustic protection -- i.e. sound-proofing, orientation with respect to the road, and insulation of the facades.

The third element, the vehicles themselves, are subject to widespread acoustic emission regulations which are necessary because of the size and international nature of the car market, despite the fact that there are notable differences in the acoustic standards of passenger cars and goods vehicles (i.e. the engine and subsidiary components). As concerns rolling noise, the interaction between tyres and pavements cannot be regulated entirely through vehicle legislation because of the great differences between the quality of road surfaces.

As regards legislation pertaining to road noise control, the more advanced countries distinguish always between the two related problems:

- existing infrastructure; and
- newly built infrastructure.

For existing roads, it is first of all necessary to define the magnitude of the problem. Actions required are then determined in relation to two different categories of noise limits, i.e. for day and for nighttime. The maximum acceptable levels are higher for existing roads compared to new infrastructure In almost all countries the limits are tied to human activities, even if the idea is taking root that many roads and their zones of impact constitute a territorial entity and thus require area-specific limits.

New infrastructure can improve the surrounding area by easing traffic flows on existing roads. There is no doubt that, at least on certain roads, the acoustic impact is essentially linked to condition of traffic congestion. Hence the construction of new roads can bring about environmental benefits through a better distribution of traffic flows in the network and the various associated transport systems. If this occurs, the noise problem must be adequately described and quantified, especially when environmental impact assessments are required for new roads. In this respect it is important to avoid unjust compensations; the benefits (reduction) obtained on the existing network should not be used to justify the acceptance of higher noise impacts of the new facility, if for no other reason than that the receptors concerned are not the same.

In all these investigations it is necessary to also evaluate the zero-option alternative.

As to long-term strategies, needless to say that it is important to implement preventive measures through effective urban planning, traffic organisation and management.

There exists a further diversity in legal approaches to the noise problem:

- Some countries have identified a corrective methodology and/or post hoc procedures to acoustically "reclaim" the zone adjacent to the infrastructure. This means that if the procedure set has been respected, the results obtained in terms of noise protection have to be automatically accepted (the "analytic" approach).

- Other countries, however, require a numerical assessment of the results obtained through the use of accepted or facultative procedures (the "experimental" approach).

Obviously the second approach is more complex. In fact to obtain the desired accuracy for noise measurements it is necessary to conduct on-site assessments which are difficult, involve uncertainty and are expensive.

From the economic point of view there are two major sources for financing noise abatement:

- fuel taxes, which produce continual revenue;
- earmarked budgets of State bodies.

These sources apply only to existing roads. Noise reduction expenditure related to new roads is internalised in the construction costs and as such is normally borne by the road authority. The two financial sources listed above may imply two different means of noise improvement:

- centralised actions for gradual improvement based on long-term programmes.
- a voluntary approach targeting specifically highly polluted areas.

To conclude, it is difficult to lay down international standards for noise abatement because the prevailing individual cultural and political values, which vary from country to country, will have to be taken into account.

TECHNICAL QUESTIONS

Major steps forward have been made in abating noise generation at source. Actions have been taken on the vehicles themselves as well as the components of the road, for example:

- *Pavements* have now reached levels of performance which would have been unthinkable only a few years ago;
- Sound insulating and sound-proofing *barriers* have become more and more efficient;
- *Combined actions* of the two types of measure have been very effective in noise abatement.

Unfortunately the relevant technologies used are costly. They require continual maintenance, as all sophisticated products, without which their effectiveness would diminish. This is particularly the case with pavements. However, as far as new roads are concerned it is now clear which path has to be followed in order to achieve accurate noise pollution control. The directions are set out in Chapter VIII.

STRUCTURE OF THE REPORT

First of all, the phenomenon of noise is defined. A minimum nomenclature of scientific terms is provided as a basis for understanding the phenomena to be controlled and the underlying laws.

A detailed picture of the legal and real constraints in OECD countries follows next, along with various criteria for assessment of not only noise but also of the relative social costs. On the basis of a majority view, acceptable limits are suggested for noise of both new and existing infrastructure, during day and night. As in other sectors of environmental impacts, it is commonly considered to be preferable to *attempt a global noise reduction over the entire territory affected rather than precisely assess whether or not the indicated limits have been reached.*

The Report presents information on how noise is measured in the various countries and addresses future potential noise developments over time and space. Methods of measurement and models for forecasting are indispensable for the design of anti-noise structures and for evaluating the results obtained.

The Study then reviews possible actions and effective technical measures that can be taken to reduce noise:

- at source, or
- in its propagation.

In the first instance the measures which could be applied to the road structure to reduce noise diffusion are discussed. Other measures which are not directly related to infrastructure -- traffic control, vehicle measures, town-planning, etc.. -- are also outlined.

Anti-noise pavements and barriers are dealt with at length, insofar as they are the most typical anti-noise road structures. These are measures that can be easily added to the road or applied by changing certain road features. In the Report they are first of all studied one-by-one and then the combined effect of their use is examined. The economic aspects of these supplementary means of protection are also assessed.

The final part of the report comprises the Expert Group's conclusions and recommendations. Here, the various anti-noise pollution approaches of the OECD countries are compared, and the overall procedure for obtaining improvements to abate road noise is presented in summary form. New roads are treated separately. The final recommendations concern rules to be adopted, focusing on noise limits, town-planning measures and road design.

Professional training and university education on the subject of noise abatement need to be enhanced.

The report closes with an outline of research requirements and current research programmes.

NEED FOR INTERNATIONAL CO-OPERATION

In order to reduce road noise there exists a considerable variety of approaches. Both legislative and methodological viewpoints play an important role. Specific technical solutions are available together with the necessary know-how. Future co-operation in this area could help in establishing and recognising a series of indicators for the most valid economic and practical measures. Such indicators would be of immediate benefit to less advanced countries but could also be of great help to other nations.

CHAPTER I

INTRODUCTION

I.1. MOTIVATION AND AIMS OF THE STUDY

Noise caused by road traffic is the nuisance the most often cited by roadside residents. Road noise is growing in OECD Member countries, as a consequence of the sustained expansion of road freight transport. Especially transport at night, is a key feature of disquiet, as highlighted in the frequent petitions by the population, for example lately by a Swiss referendum that has approved a ban, to be enacted within 10 years as of 1994, on the transit of trucks on main roads.

The following table presents a French prediction for the year 2010 of the changing proportions of the urban population exposed to different noise levels.

Table I.1. Exposure of the urban population to road noise in France (measured at the building fronts)

Leq 8 h-20 h (on front face of buildings)	1985		2010- (estimated)	
	%	Million inhabitants	%	Million inhabitants
< 55 dB (A)	46.4	17.0	49.5	18.8
55-65 dB (A)	37.2	13.7	40.2	15.2
> 65 dB (A)	16.4	6.1	10.3	3.9

Noise is not the only type of nuisance by road traffic. An in-depth study of environmental road impacts has been recently carried out by the OECD Road Transport Research Programme and published under the title "Environmental Impact Assessment of Roads" (1994).

Noise reduction arising from the implementation of motor vehicle standards that improve the acoustical qualities of vehicles is very slow. Also, regulations concerning mobility and speed have negative consequences in general economic terms, and are difficult to enforce. It is therefore necessary to implement in the short term various noise reduction techniques focusing both on planned and existing road infrastructures, the latter being the source of the main problems to be solved.

Noise problems will be increasingly perceived by populations living in the proximity of roadways. However, as of today, there are few citizens in OECD countries that are willing to spend money "directly" to reduce this nuisance, which is usually considered to be a problem to be resolved by "others": the vehicle manufacturers, the owners or operators of roads, the State. This perception may vary from country to country given the diversity of regulations, limits and financial systems, as discussed in Chapter II of the present report.

The importance attributed to the noise problem varies with individuals and communities. Therefore it is interesting to note the different approaches to noise limits and levels:

- Some countries have very severe limits, based on medical criteria, considering that an area or zone can be managed using the same criteria as for the interior of a building, and that prolonged exposure to traffic noise has highly pathological effects;

- Others, more pragmatically, regulate in a reasonable fashion the global sound energy affecting the area to be protected over a given lapse of time.

It should be noted that overly severe regulations do not always result in effective protection measures, since it is technically impossible to attain the established levels; whereas regulations that may seem to be more permissive may actually assure widespread application of noise containment programmes in the vicinity of roads. All this amply justifies the international cooperation that has been at the origin of this study which after having analysed practices and research in different countries, sets out options for methods of measurement and limits that are technically acceptable. Many of the levels of standards proposed or imposed by various countries are not attainable in practice even with very high capital investments.

By indicating measurement criteria -- measurement units, timings, and sites -- it is possible to clarify the evaluation issue of noise, even for non-specialists. Noise control methods can then be suggested (with their relative costs), both specific measures -- such as low-noise pavements and anti-noise screens -- as well as larger actions pertaining to the whole road structure.

I.2. OUTLINE OF REPORT

The main function of the Expert Group was to gather information mainly on the techniques that inhibit the propagation of noise, such as *antinoise screens* or *noise barriers*, and on those related to both the emission and the diffusion of noise, such as *pavements,* as well as *combinations of both systems.* The IRRD[1] key words relate to noise; low noise pavement; pavement; noise barrier - and also to noise measurements, noise limits and the mathematical modeling of noise. The Expert Group introduced two main topics into a larger and more articulated context, which clarifies the complex phenomenon of road noise, including its evaluation and its abatement.

---

1 IRRD: International Road Research Documentation of the OECD

The reduction of roadside noise is not tied exclusively to special technical means such as low noise pavements and noise barriers, whether of a reflective, isolating or absorptive nature, but also to a series of other actions that can be grouped under the following headings:

- criteria for the regulation of acoustical emissions;
- measures concerning vehicles and traffic;
- measures to be applied to roadway features other than pavements and screens;
- measures related to roadside buildings and areas.

Hence after having examined the regulations and limits prevailing in the different OECD countries (Chapter II), Chapter III provides the criteria to be used in evaluating noise, both in terms of prediction (noises that do not yet exist), and in terms of measurement. The report then includes elements that discuss all of these actions (Chapter IV), pointing to potential and possible synergies.

The survey of methods used for predicting and measuring noise is one of the most complicated aspects for non-specialists: an analysis of the data provided, together with available bibliographies, provided valuable criteria for the future understanding of the problem, depending on the options chosen. Briefly, as stated in Chapters II and III, there are three different approaches to reducing traffic noise:

- Some countries control noise through regulations which evaluate noise using predictive mathematical calculations; they apply selective remedies without generally conducting precise numerical measurements, since these would always be variable depending on the conditions and the context at the time of measurement. (Netherlands, United States, Norway).

- Other countries, instead, follow a more numerical orientation, closely tied to measurements, and use prediction models (France, Germany, Italy, Spain).

- Other countries follow intermediate solutions.

All systems are equally valid, although it would seem that the most effective results can probably be best obtained by integrating the methods. The sound is not such that it can be measured as an absolute entity, but has to be qualified by specifying the location of the measurement as well as the prevailing conditions. Data for the criteria to be applied are provided in Chapter III which discusses at length the methods with respect to their objectives and efficiency.

As mentioned, Chapter IV provides information on non-specialised techniques for noise abatement. This is the most important chapter for the design of new infrastructures, since these constitute the basis for the most rational means of reducing noise.

Different effects can arise from roadway designs (tunnels, cuttings, embankments and viaducts), but one should also bear in mind the periphery including vegetation. Protection measures that can be applied to adjacent buildings must be considered, especially when the cost of installing protection systems along the roadway would be higher than that of controlling the reception capacities of buildings.

Chapters V and VI review some special types of measures:

- low noise generation and/or noise-absorbent pavements;
- protective screens made of different materials and having different structures.

The use of special pavements has reached a maximum with the application of noise-absorbing porous (drainage) asphalts and low emission thin layer wearing courses, but many problems still exist with respect to pavements that need to be studied further:

- the different treatment of urban and non-urban roads;
- the classification, in acoustical terms, of different anti-noise materials;
- problems relating to performance measurement;
- maintenance systems and materials used (acoustical effects over time).

For noise barriers the range of information is even broader and the study has attempted to provide classification criteria related to:

- materials;
- acoustical efficiency (acoustical isolation/absorption);
- construction systems;
- visual impacts.

Chapter VII is devoted to the evaluation of the effects of different solutions taken together, providing indications, based on the information collected, on the different problems related to specific solutions and providing data for their optimisation -- i.e. compatibility with each other, avoidance of duplication effects and limitations to costs. This Chapter also discusses costs, assembling the data provided in other parts of the Report. It is possible to achieve a "quiet" road integrating the different technical systems, but the results must be analysed each time, because it is not always possible to simply add the effects of all these actions.

The discussion on research (included in Chapter VIII) has attempted to highlight the aspects common to both types of abatement measures, and provide information helpful for a widespread use of the potential treatments. Current research is linked to the different approaches to noise control, i.e.:

- There are countries that give priority to the sociological aspects of noise (generally the Nordic countries) with research emphasis on the effects of acoustical conditions on people;

- There are countries that search for solutions related to the physical aspects of the problem, and which study in numerical terms how to measure, control and evaluate noise.

Several conclusions and recommendations arising from the report are put forward.

GLOSSARY*

The dB(A) -- an adjusted unit of measurement

While the level of a sound presumed to be constant is correctly expressed in decibel (dB), it is necessary to point out that legislation in force in several countries uses different types of decibel, depending on the noise sources considered. The noise generated by road and rail vehicles is measured in "A" decibel units (dB(A)), the "A" meaning the level of noise registered by a microphone that is filtered and adjusted in the same way that the human ear filters and adjusts the sounds it receives. It is necessary to add that this preference is well justified for noise of medium-low and high frequencies; to take into account the effects perceived by a human being above 50 dB, and in the presence of freight vehicle, a different type of filtered curve is coming into use (C curve), therefore one will find the expression written as dB (C).

Addition of decibel measurements

Sound units of measurement are more difficult than, for example, the better known units of linear measurement: while it is possible to directly sum two lengths, the same is not possible for two noise levels. The simple addition of two decibel levels is impossible, because the decibel scale is logarithmic and not linear. It is necessary therefore to bear in mind that the sum of two sounds of the same level produces a total level such as 70 dB(A) + 70 dB(A) = 73 dB(A). If the second noise level is lower than the first, the result is : 70 dB(A) + 60 dB(A) = 70 dB(A).

An acoustical index to define discomfort: the Leq and LAeq

The sounds generated by vehicle circulation are fluctuating; it is therefore necessary to be able to characterise them in a simple manner in order to predict the level of discomfort to those adjacent to it. For this purpose the Equivalent Energy Level, or Leq, is used: this is the presumed constant level of acoustical pressure,in which the quantity of acoustical energy emitted during a defined period would be the same as that of the effective, fluctuating, noise. There can be a Leq of a minute, an hour, a day and so forth. (see figure I.1, where the noise level can assume different values. If the curve is filtered in ponderation A, it will be written as LAeq(...). The number in parentheses that follows eq indicates the time t over which the equivalent sound level has been measured. Other sound dimensions are Lmax (see fig.) that indicates the noise level for an isolated and low duration event. Lastly there is L10 and L50, that indicate the sound level exceeded for more than 10 or 50% of the considered time.

Figure I.1. **Representation of LAeq**

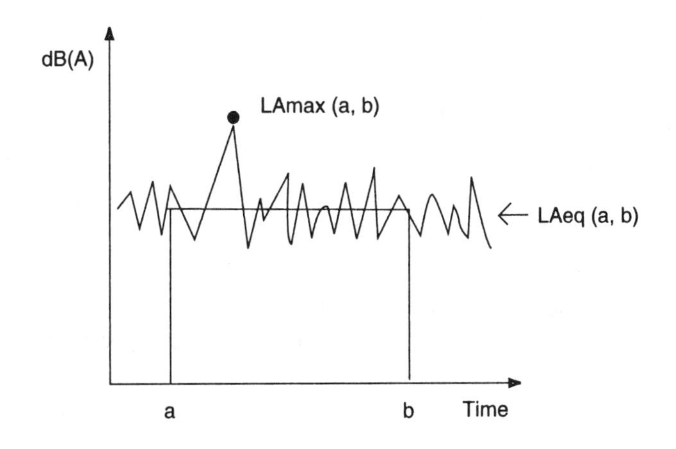

* Source : J. LAMBERT (INRETS, France)

Table I.2. **Examples of noise levels in dB(A)**

Acoustical Pressure (Pa)	**Level of Sound Pressure [dB(A)]**	**Acoustical Sensation**	**Examples**
< 2.10^{-5}	<0	Not audible	Anechoic room
2.10^{-5}	0	Threshold of audibility	Audiometer tests
$6,3.10^{-5}$	10	Very quiet	Recording studio
2.10^{-4}	20		Caves, undisturbed snowfields
$6,3.10^{-4}$	30	Quiet	Bedroom
2.10^{-3}	40		Quiet office
$6,3.10^{-3}$	50	Moderate	Office
2.10^{-2}	60	Bothersome (for intellectual work)	Speech level at 1 meter
$6,3.10^{-2}$	70	Moderately loud	Pedestrian road, tailor shop
2.10^{-1}	80	Loud	Train station
$6,3.10^{-1}$	90	Damage threshold if >8 hrs per day	Modern machine tool workshop
2	100	Very loud	Machine shop, glassworks, steel lathe machinery, weaving machinery
6,3	110	Shouted words are inaudible	
20	120	"Deafening"	
63	130	Pain threshold	Plane jet engine, motor test group

The indicated levels are acoustical pressures (Lmax). The road noise level, measured in LAeq, in urban areas falls in the range of 55 to 75 dB(A). Country side in winter night have a noise Level in LAeq of 35 to 40 dB(A).

CHAPTER II

CURRENT PRACTICES AND NOISE LIMITS

II.1. INTRODUCTION

In this chapter the general policies, which are currently adopted in OECD countries for controlling traffic noise, have been examined mainly as regards their normative aspects. All the information has been interpreted on the basis of an homogeneous judgement, in order to get common guidelines which could be used as the basis for OECD recommendations.

Particular attention has been paid to the following points:

- Parameters used to describe the exposure of populations to traffic noise and to fix the allowable limits (Section II.2);
- General criteria on which standards are based (Section II.3);
- Economic and financial management of anti-noise policies (Section II.4).

II.2. EXPOSURE OF POPULATION AND ALLOWABLE LIMITS

As is evident from the analysis performed, and from the conclusions of other OECD documents (1), road traffic noise still continues to be considered as a major environmental problem; moreover, from a general point of view, the most up-to-date studies seem to indicate only a very slight improvement, in comparison with the situation 10-15 years ago.

As regards the measured levels -- and when one considers the subjective or social aspects -- things appear more complex: for example the last INRETS (2) research shows that in France the percentage of the population which considers itself seriously annoyed by traffic noise has grown from 25 per cent in 1976 to 31 per cent in 1986. An Italian study (3) points out that in 1992, 22 per cent of the population considered traffic noise an urgent problem to be resolved, while the percentage in 1986 was only 10.6 per cent. In contrast, the most recent results show that in The Netherlands an appreciable improvement has been obtained along motorways, whereas along city roads the situation seems to be unaltered, as shown by the figures in table II.1. Of course these results take into account the anti-noise actions (barriers, insulations, porous pavements) as well as the growth of traffic volume. Denmark too has experienced similar results.

Table II.1. **Evolution of the annoyance due to traffic noise in The Netherlands**

	Annoyance score (%)		Score seriously annoyed (%)	
	1988	1993	1988	1993
Motorways	59	47	21	12
City roads	46	48	12	10

In any case, from all the research and studies carried out until now, it appears clearly that traffic noise, at typical levels of emission, does not cause any immediate risk of hearing loss. Road noise has, however, been proved to cause important non-auditory effects, which can be grouped in the following manner (4):

- Disturbance of activities: firstly, those concerning communication; secondly those involving work and concentration; thirdly, those concerning rest and relaxation;

- General annoyance, that is a condition of psychological disturbance, which can sometimes lead to psychosomatic disturbance, such as stress or psychiatric disorders when the individual is particularly sensitive (5).

Figure II.1. **Road traffic noise annoyance in France**

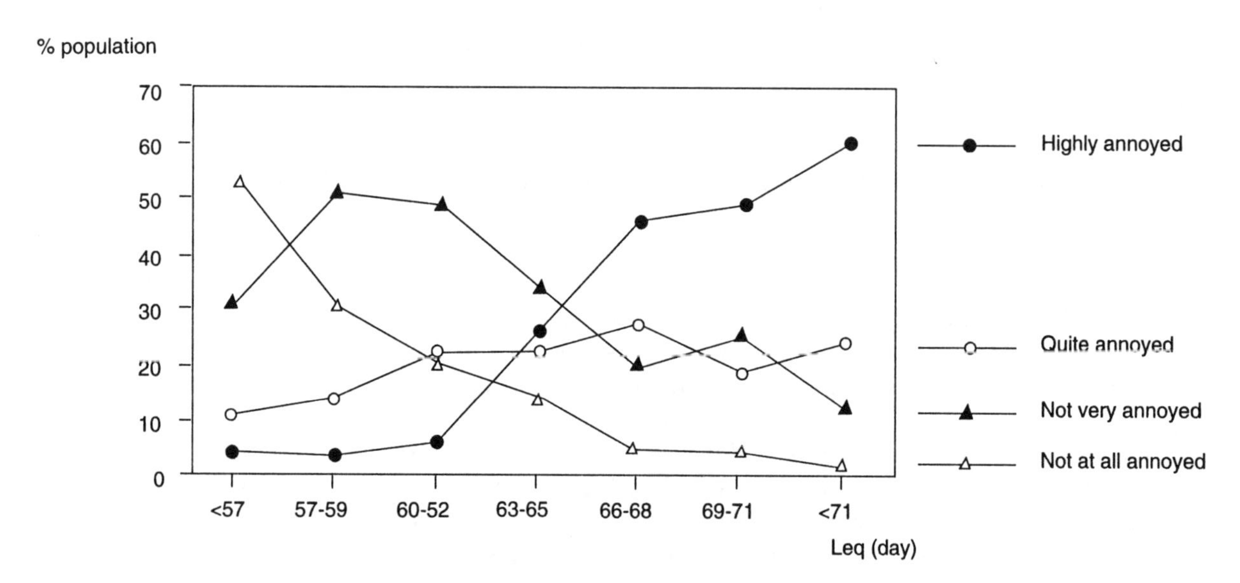

Figure II.2. **Road traffic noise annoyance in Modena (Italy)**

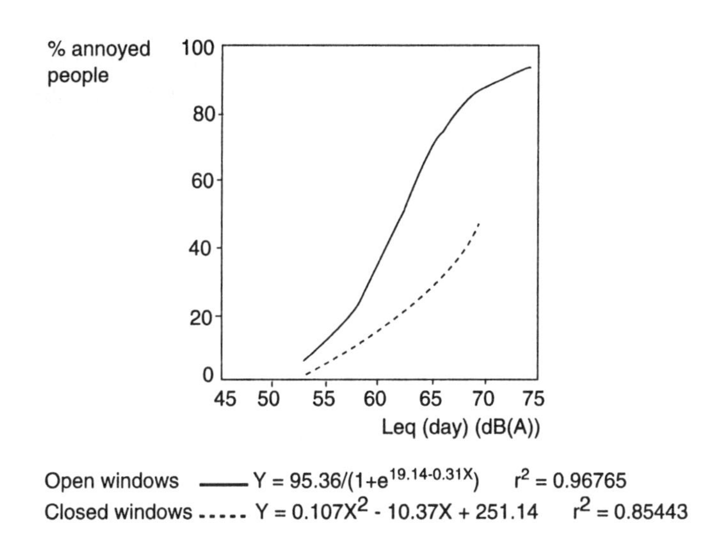

Of course all the noise control policies refer to these effects, and from a general point of view, one can state that the disturbance of activities is controlled by means of interior limits (i.e within buildings), while annoyance is controlled by exterior (facade) limits.

It is therefore obviously important to choose properly the parameters used to describe disturbance and annoyance. In fact, firstly, if the chosen parameters are truly representative of human susceptibility, consequently all the anti-noise actions will be positively perceived and supported by the population; secondly, it will be possible to fix noise exposure limits by balancing economic and technical aspects versus a threshold of discomfort that has been previously examined and fixed (this latter point will be further discussed in section II.3, where economic and financial aspects will be touched upon).

After a review of the situation in various OECD countries, the following conclusions can be drawn:

- The "A" filter is the most frequently used weighting curve for environmental assessment. The characteristic of "A" weighting is that it corresponds to the reduced sensitivity of the human ear to frequencies above 8 Hz and below 200 Hz at a relatively low loudness level of 30 phon, i.e. where the standard equal loudness contour passes through 1000 Hz at 30 dB(A).

- The equivalent continuous sound level, Leq, is the level most frequently used to describe "disturbance": all the countries which take into account interior limits (inside rooms), refer to values expressed in Leq. (The equivalent continuous sound level is defined as the level of that (hypothetical) constant sound that, over the period of measurement, would deliver the same sound energy as the actual intermittent, or time-varying sound.)

- As regards "annoyance", as previously stated, it is referred to outside facade levels. In this case the exterior situation is not as homogeneous as in the case of disturbance. Alongside many countries that use Leq, there are in fact many where L10, Lmax or L50 are used (United Kingdom, Australia, New Zealand, USA, Norway). [The statistical levels (L10, L50) are defined as the sound level in dB(A) exceeded for 10 per cent and 50 per cent of the measurement period respectively].

- ♦ In general traffic noise is compared only with threshold values; exceptions are to be seen in Italy and Australia, where the "differential criterion" is used, that is to say the comparison is made between the actual road traffic noise and the ambient (or background) noise.

- ♦ A clear discrepancy is observed regarding the reference to time intervals during which noise is assessed: roughly 60 per cent of countries have two values, daytime and nighttime, with a common constant difference between daytime and nighttime figures of 10 dB(A). The others use a single value, differently expressed. It is interesting to note that, due to typical road usages, the adoption of the two-sided limits leads frequently to redundant imposition or to anomalous situations.

  This issue will be the subject of further psycho-acoustical research, for a better understanding of the phenomena involved (see chapter VIII).

In conclusion to this paragraph, one can refer to the final statement of a recent study (5) carried out by INRETS for the Commission of the European Communities: "Our proposal is to use the Leq to describe noise, and to use additional weighting in some specific situations, depending on sources, and the nature of the noise". As a result of this concept, a new study is now being undertaken in Italy on the use of the "C" weighting scale for road noise in the case of a consistent density of trucks or marked stop-and-go conditions.

II.3. GENERAL CRITERIA FOR STANDARDS

Two different policies seem to be adopted by OECD countries for the abatement of traffic noise: a "voluntaristic approach" versus "long-term centrally controlled action".

The first approach consists of the decentralisation of activities, sporadic repair of the most critical situations (identified mainly by private complaints), public information campaigns to promote correct behaviour, incentives to car manufacturers to develop quieter vehicles, incentives and control of the market (both economic aids and concession of a greater freedom of use for quiet vehicles). Countries such as Australia, France ("Villes pilotes" and "Rattrapage des points noirs"), Italy (1989-92 DISIA programmes of the Ministry of Environment), Germany, Spain and the USA belong to this category.

The second approach involves mainly long term plans, aimed at a general noise exposure within preset, allowable limits: these programmes are usually managed by some Central Authority, and they provide for measures both on roads and on buildings (noise barriers and windows). Countries such as Austria, Denmark, Finland, The Netherlands, Norway and Switzerland belong to this category. Examples of this approach are the Danish Action Plan on Transport for the Environment and Development [100.000 homes with less than 65 dB(A) by the year 2010], the Austrian Parliament Statement [65 dB(A) daytime and 55 dB(A) nighttime, to be reached for federal roads only, by the year 2003], the Dutch Second Structural Scheme for Traffic and Transport [in the year 2010 the number of homes with no more than 55 dB(A) at the facade shall be 50 per cent of the 1986 situation] and the Swiss Federal Ordinance of 1986 (observance of the legal threshold values by the year 2002).

It is really difficult to prove in practical terms which method is the more effective: generally speaking one can only observe how the "voluntaristic approach" can often lead to an adapted "in-situ" procedure, whereas the "planning approach" forces the promotion of major actions which have great economic impact.

Moreover, when the "planning approach" is adopted, a substantial amount of funds is dedicated for the required actions: these resources are generally obtained by means of fuel taxes (Switzerland and the Netherlands), in order to have a revenue that is more or less proportional to traffic noise levels. On the contrary, the "voluntaristic approach" is generally based on the state budget, which can vary from year to year, depending on economic or political considerations. It appears that anti-noise regulations and policies are better respected and implemented when a specific budget to reduce noise is allocated, and when the procedures for obtaining proper financial resources are established.

By examining the situation in the various OECD countries, one can note how the limits are always diversified according to typology of land use or human activities: categories range from a minimum of 3 (i.e. Germany, for existing roads and buildings) to a maximum of 18 (Italy, for different types of roads, with existing buildings). One important point is that a few standards (Italy, Japan and, to a certain extent, the Netherlands) recognise the road together with a portion of the surrounding area as a separate category, to which specific limits are assigned: in this case, roads are classified according to the number of lanes, or to the rate of traffic flow.

In the Italian case the "road zone" has a fixed width. In the Japanese model the area is identified by the buildings facing the road, while according to Dutch rules the width varies according to the number of lanes and the land use (rural or urban areas).

One interesting and original approach is used in Australia, namely in New South Wales and Queensland where, for new roads, the initial noise environment is used to calculate the limit not to be exceeded.

No explicit distinction is made between motorways, highways, inter-urban roads, urban roads (small, medium and large towns), ring roads and so forth; perhaps this point should also be taken into account in the future, considering the substantial differences existing between these kinds of roads. The applicable technical measures are also quite different, depending on the place where they are applied: in densely built-up urban areas for example, barriers and earth berms cannot be easily used, whereas traffic management is easier to perform; this situation is completely reversed on motorways and highways.

Almost all of the existing standards make a substantial distinction between existing or non-existing roads and buildings (priority criterion), by assigning a proper threshold value to the various possible combinations; limits are quite different, ranking from a minimum of 5 dB(A) to a maximum of 15 dB(A). The general approach for new roads or new buildings is toward "mandatory standards" (action is compulsory), whereas for existing roads and buildings standards act as "guidelines" (action is not compulsory, exemptions can be obtained, limits are considered as tendencies to be achieved).

To establish whether an action must be undertaken, the allowable limit is compared to the "actual noise level". Experimental (measurements) and analytical (calculations) procedures can be used to evaluate this actual noise level. Both methods present advantages and disadvantages, summarised in table II.2.

For a detailed discussion, please refer to chapter III.

Table II.2. **Comparison of experimental and analytical procedures**

	Experimental procedure	**Analytical procedure**
Advantages	• Good estimate of real conditions • Objective results • Final inspection of results	• Easy standardisation • Reasonable cost • Easy to perform
Disadvantages	• High costs • Time consuming • Need of expert personnel • Influence of external atmospheric parameters • Applicable only in pre-existing contexts or in laboratory models	• Estimate of ideal situation • Uncertain accuracy

Even if it is not very explicit, it is possible to identify a sort of common approach to fix allowable limits. First of all it seems that almost every country has carried out psycho-acoustic and socio-acoustic research in order to correlate as much as possible disturbance and annoyance with measured parameters. At this stage the limits have been fixed according to the following criteria:

For nighttime values: an internal noise level that does not disturb sleep has been chosen in the range of 35-40 dB(A). This value has been maintained as is, or in some cases it has been "projected" outside the facades, taking into account the insertion loss of closed windows. It is important to note that this situation is peculiar to road traffic noise while, for example, Lmax or L10 are frequently used for railway noise.

For daytime values: a percentage of people "highly affected" by traffic noise has been chosen between 10 per cent and 25 per cent, and the corresponding continuous equivalent level in dB(A) has been set as an allowable value on the facade. (see figure II.1 and II.2 -- the French and Italian studies). This level has been kept as a "basic" value for medium quality areas, and in addition, two or more values have been established for low and high quality areas. Later on, the "priority concept" has been applied to classify situations of existing roads and buildings.

II.4. ECONOMIC AND FINANCIAL ASPECTS

The economic and financial aspects are of great importance, since they can influence the final results of anti-noise policies. It is also evident that a correct noise reduction policy must necessarily take into consideration a cost/benefit analysis. When speaking of "costs", two specific elements must be considered:

- The cost of prevention and remedy, targeted at bringing the level of noise pollution within acceptable standards;

- The social cost paid by the population, linked to a policy of non-intervention on the part of the Authorities.

In the following paragraphs both of these topics will be discussed, beginning with the analysis of the costs relating to possible actions to reduce traffic noise.

When adopting laws aimed at controlling and reducing road traffic noise pollution, it is important to examine beforehand the related economic implications, taking into account all the solutions which are technically feasible.

Considering that WHO (World Health Organisation) suggests daytime values of 65 dB(A), and that the actual levels caused by road traffic in noisy areas are in the range of 70-75 dB(A), the gap is between 5 and 10 dB(A). To obtain this noise reduction, it is possible to intervene on:

- *Sources* by means of new, and more silent vehicles, and by the improvement of existing vehicles;
- *Infrastructures* through the use of noise barriers, low-noise road surfaces, acoustical ceilings, tunnels, etc.;
- *Buildings* by improving the soundproofing of facades, roofs and especially windows.

II.4.1. Actions on vehicles

Concerning vehicles (passenger cars, buses and trucks), reference can be made to the results of a specific study promoted in 1983 by the Commission of the European Communities (8) (9). It appears that the extra cost necessary for reducing noise emissions by between 5 and 10 dB(A), considering the type approval test, is:

- between 2 and 5 per cent for passenger cars;
- between 5 and 9 per cent for trucks.

When we apply these figures to the road vehicle stock in OECD countries (year 1991; 353 million passenger vehicles and 93 million goods vehicles, mean sale prices of 15.000 and 50.000 US Dollars respectively), considering that a complete turnover of vehicle stock will occur within the next ten years, the figure comes to an extra cost of US$ 51 billion per year. This is of course a very rough estimate, but it is nonetheless useful for considering the question.

II.4.2. Actions on roads and buildings

Similar analysis can be performed by taking into consideration the actions necessary for roads and buildings.

If one uses the data provided in references 10, 11, 12 and 13, the following figures can be obtained, taking into consideration only the improvements (existing contexts) for a daytime limit of 65 dB(A), and for Germany only, of 65 dB(A) daytime and 55 dB(A) nighttime limit:

Germany	(all roads)	=	60 000 US$ /km
France	(all roads)	=	27 000 US$ /km
Netherlands	(roads in built up areas)	=	16 000 US$ /km
Switzerland	(all roads)	=	28 000 US$ /km

It should be noted that the data reported in references 10, 11, 12 and 13 have been revised to apply to 1994 with a mean inflation rate of 3 per cent. All of the data presented are estimates based on theoretical and statistical elements; a subsequent evaluation, derived from the Swiss programme for the improvement of national roads, gives a cost of between 42 000 and 59 000 US$ per km.

Taking the mean value, 35 000 US$ /km, and the combined road network length of all OECD countries, the following figures can be obtained:

All roads in the OECD	437 billion US$
All motorways in the OECD	4.8 billion US$

As with the case of costs related to vehicle improvement, it is important to emphasize that the figures presented thus far are only targeted at providing a rough idea of the "size" of the problem, as the following points must also be taken into consideration:

- Estimates are based on studies carried out at the end of the seventies, and they do not take into account technical progress which can reduce the cost of anti-noise actions: by looking at the table in chapter VII for example, it is possible to see how much the situation could differ with the adoption of traditional and new low noise surfacings (euphonic or optimised).

- Each country has specific characteristics that can influence costs enormously: in cold-climate countries for example, it appears difficult to adopt porous pavements in view of ice related problems, while a policy for improving window insulation properties could be accomplished in the meantime through energy savings programmes; this situation is present in exactly the reverse context in warm climate countries.

It is also necessary to underline the different situations regarding the density of population around the road network, and the predominant size of urban areas.

Taking into account all factors, it can be stated that the mean cost of 35,000 US$ per km can vary in a range of +/- 50 per cent.

Even with this limitation in considering the "low cost" figure, it is evident that a realistic anti-noise policy requires long term planning; in Switzerland for example, the Programme for Roads will end in 2002, and in the Netherlands the Noise Abatement Act schedules activities until 2010.

Another very interesting point is to examine to what degree the investments in anti-noise policies are influenced by the levels assumed as goals (or warning thresholds). According to the references 10, 11, 12, the figures shown in tables II.3, II.4 and II.5 are valid. Even if these figures are to be considered only as approximate values, it is evident that the choice of the limits can greatly effect the total economic impact of anti-noise measures.

Facing this consideration, one can note that the studies aimed at correlating the annoyance to the noise level seem to be neither very accurate nor up-to-date. This leads to considering traffic noise as a homogeneous problem, without making a distinction, for example, between urban roads (main and secondary), and motorways, or between free-flow and stop-and-go traffic conditions.

Perhaps more accurate investigations could highlight differences between countries, as well as between varying sensitivity to the various types of traffic noise. Most of the standards seem to be based on criteria derived from psycho-social research carried out 15/20 years ago. It could be

Table II.3. **Estimated cost of noise level abatement for existing roads in Germany (1994)***

Limit Leq(day)/Leq(night) dB(A)	Cost (Million US $)
80 / 70	473
75 / 65	3350
70 / 60	10645
65 / 55	30975

* Updated from the 1980 values (13) and related to 486.000 km of roads

Table II.4. **Estimated cost of noise level abatement for construction of new roads in Germany (1994)***

Limit Leq(day)/Leq(night)	Cost (Million US $)
65-70-75 / 55-60-65	485
60-65 / 50-55	858

* Updated from the 1980 values (4)

Table II.5. **Estimated cost of noise level abatement for all roads in Switzerland (1994)***

Limit Leq(6 a.m.-10 p.m.) dB(A)	Cost (Million US $)
75	149
70	683
65	962

* Updated from the 1979 values (10)

interesting to check whether the conclusions drawn in the seventies can still be considered valid today, or if any changes must be taken into consideration (such as a more intense sensitivity to noise pollution or the opposite, i.e., a trend of being more accustomed to noise pollution).

One should also consider whether noise levels or peak noise levels are better indicators of noise disturbance, or whether one would need a genuine traffic disturbance measure, taking into account that people experience traffic as a whole, including not only noise but air pollution, traffic safety, visual intrusion or obstruction. The Norwegian "Traffikk og miljo" for example, indicates support for a holistic view.

II.4.3. Social cost of noise

Another factor to be considered is the "social cost" of noise, i.e. the price paid by the population as a result of non-intervention policies. It is obvious that in this second case the studies and analyses are even more difficult than before, as it is necessary to evaluate, in economic terms, the damage caused by noise to both material goods (real estate, parks, building areas) and to persons (stress, illness, reduction of worker productivity.

The most commonly used evaluation methods are:

- Contingent evaluation, or the analysis deduced from statistical investigations on how much individuals would be willing to pay in order to benefit from a better acoustical environment;
- The analysis of the influence of acoustical pollution on certain market indicators, mainly based on the values of the real estate market (purchase/sale and rental of properties);
- The study of legal history, analysing the judgements through which courts have established indemnities for those who have suffered acoustical nuisance or damage;
- An analysis of the expenses incurred by individuals for the acoustical protection of building facades.

By synthesizing the results of the most authoritative studies carried out (14)(15)(16)(17)(18), it is possible to deduce the following general considerations:

- In Germany the average payment consensus to obtain a quieter environment is equal to 1.2 US$ for each dB(A) of reduction, for all types of noise. Considering only the road noise factor, the total estimated cost for the German market would be of the order of 0,75-0.85 billion US$ per year.
- By analysing the studies carried out between 1980 and the present, one obtains an average annual depreciation rate of real estate of 1 per cent. On the basis of similar data, it has been calculated that in France, homes located within the urban areas of cities with more than 50.000 inhabitants have suffered an evaluated loss of 20 billion US$, assuming that depreciation begins at daytime levels equal to 60 dB(A).

It is necessary in this case as well to emphasize that the estimates carried out must be considered broad approximations. In spite of this, it is possible to underline that the order of magnitude of the damage suffered is similar to that of the methods of amelioration. This point is of vital importance for the correct planning of strategies to combat noise pollution.

II.5. CONCLUSIONS

The following basic suggestions can be proposed:

- LAeq can be considered as the "basic" descriptor of the exposure of populations to road traffic noise, but further investigations are needed in particular conditions. For example when the percentage of heavy vehicles is high, it could be useful to apply the C weighting filter; when the volume of traffic is very low, Lmax or L10 could be useful;

- The correlations between on the one hand annoyance and disturbance and on the other the measured or calculated describing parameters have to be well known before proposing permissable noise limits, as the economic and technical implications are very important;

- Anti-noise policies need long-term plans and a well defined financial approach to obtain adequate funding;

- An integrated approach (vehicles, roads, barriers, buildings) is necessary to provide realistic technical solutions and to control expenditures;

- The priority concept has to be carefully considered, by making a distinction between:

  - construction of new roads, or reconstruction of roads near existing buildings;
  - construction of new buildings or reconstruction of buildings near an existing road;
  - existing roads and buildings.

Taking into account the current level of technical knowledge, and the economic implications of compulsory policies for the control of traffic noise, it is be possible to suggest some general indications related to the permissable limits in the medium term -- by 2005-2010 -- see table II.6:

Table II.6. **Proposed Leq levels**

ACCEPTABLE LEVELS (Facade Limits *)			
LAeq, daytime		LAeq, nighttime	
New road	Existing road	New road	Existing road
60 +/-5	65 +/-5	50-55	55-60
* If the facade limit is not technically attainable, or if its being attained is not economically justifiable, it is necessary to guarantee a closed-window internal level of 40/50 LAeq(1h) during the daytime and of 35 LAeq(1h) during the night			

The proposed OECD limits are strongly coherent with the figures expressed as "satisfactory" by the Commission of the European Communities (DGXI) as shown in table II.7.

Table II.7. **Scenario to provide satisfactory protection to people exposed**

LAeq (Facade limits)			
LAeq - Daytime		LAeq - Nighttime	
New installation	Existing installation	New installation	Existing installation
57/68	65/70	47/58	57/62

II.6. REFERENCES

1. OECD (1991). *Fighting Noise in the 1990s.* OECD, Paris.

2. MAURIN, M., LAMBERT, J. and A. ALAUZET (1988). *Enquête nationale sur le bruit des transports en France.* Rapport INRETS n°71, Bron.

3. IPA - CENSIS (1993). *I comportamenti Ambientali Autostrade 3/93.* Rome.

4. BERTONI, D., FRANCHINI, A. and J. LAMBERT (1994). *Gli effetti del rumore dei sistemi di trasporto sulla popolazione.* Pitagora Editrice. Bologne.

5. LAMBERT, J. (1994). *Study related to the preparation of a communication on a future EC noise policy.* LEN Report N°9420, INRETS. Bron.

6. MINISTERIO DELL'AMBIENTE (1988). *Programma triennale per la tutela ambientale 1989-1991.* Ministerio dell'ambiente. Rome.

7. WORLD HEALTH ORGANISATION (1980). *Environmental Health Criteria 12 : Noise.* Genève.

8. COMMISSION DES COMMUNAUTES EUROPEENNES (1983). *Perspective de réduction du bruit des véhicules routiers.* Rapport EUR 8573 EN. Bruxelles.

9. CCMC (1990). *Estimation of the economic and technical consequences of a further reduction in permissible motor vehicle noise levels.* Rapport N° 20/80. Bruxelles.

10. DEPARTEMENT FEDERAL DE L'INTERIEUR (1979). *Valeurs limites d'exposition au bruit du trafic routier.* Commission fédérale. Berne.

11. MINISTERIE VAN VOLKSHUISVESTING (1988). *Noise Abatement in the Netherlands.* Ministerie van volkshuisvesting. La Haye.

12. BAR, P. (1979). *Protection contre le bruit et la pollution - Coût des protections en France.* CETUR. Bagneux.

13. CETUR (1981). *Coopération franco-allemande dans le domaine routier.* Rapport CETUR 1981. Bagneux.

14. LAMBERT, J. (1986). *Nuisances sonores et coût social de l'automobile.* Revue Recherche - Transport - Sécurité n°11, INRETS. Bron.

15. IRT/CERNE (1982). *L'impact du bruit et de la pollution émis par la circulation automobile.* Rapport N°41, IRT. Bron.

16. WEINBERGER (1992). *Gesamtwirstschaftliche Kosten des Lärms in Deutschland.* Larmbekampfung 39.

17. ALEXANDRE, A. and J.P. BARDE (1987). *Transportation noise reference book.* Paul Nelson (ed.). Butterworth and Co Ltd. London.

18. BOURDIN, D. (1992). *Coût social du bruit en Allemagne.* Echo Bruit. Neuilly sur Seine.

CHAPTER III

ASSESSMENT AND MEASUREMENT

This chapter provides an overview of the traffic noise assessment methods currently used in OECD Member countries.

Measurement and prediction methods must be used to quantify noise impact in all stages of the decision-making process, from the very start of the planning process to the final detailed design of anti-noise measures. The decisions taken in the course of the assessment process depend on the results provided by these methods.

Traffic noise levels can be evaluated by two different means: measurement and prediction. *Measurement methods* use acoustical instruments such as sound level meters to make direct measurements of noise. *Prediction methods* are based on acoustical theories of sound emission and propagation, which are used to calculate noise levels by simulating real or predicted situations by means of mathematical or physical models. Frequently, measurement and prediction methods are combined to provide a better or merely a more operative assessment. In current practice, two features determine the quality of a method:

- its validity, which means the accuracy of the results it provides; and
- its operativeness, in terms of both time and economic costs.

III.1. SCOPE AND USE OF MEASUREMENT AND PREDICTION METHODS

The quantitative evaluation of traffic noise levels is the basis on which noise control policies stand. Assessment tools are needed to establish existing noise levels, evaluate the impact of traffic noise in the planning processes, and determine the effectiveness of anti-noise measures.

Measurement methods are only relevant when applied to existing situations, whereas prediction methods can be used for both existing and planned situations.

From a technical view-point, prediction methods are better for determining the sound level due to road traffic. Their lower cost and greater reliability suggest that they should be preferred when assessing noise. In fact, when a calculation method is used, a large number of scenarios can be generated by introducing different traffic flows, several types of pavement, variable numbers and positions of reception points, and different noise abatement measure designs.

By contrast, measurement results give information only about a very limited situation - the specific conditions at the time the measurements are made. Traffic and weather conditions vary with time, so only strictly simultaneous measurements can be compared, unless corrections are made. Moreover, it takes some time to measure a relatively small set of points, whereas calculation methods define noise levels quickly for extensive areas.

Measurement methods are mainly used to determine noise levels prior to the construction of roads, so as to allow the prediction of future increases of level. The measurement of traffic noise may also be carried out when standardised calculation methods are unlikely to provide acceptable results: they are especially unreliable when the single noise effects involved are not known (such as emissions from certain vehicles and some types of pavements and surfaces). Moreover, complex conditions of noise propagation are not considered in calculation methods, which tend to simplify propagation scenarios.

Measurement is also used to determine the effectiveness of noise barriers. The noise levels before and after the barrier is installed are compared. Measurements are also made to test the accuracy of predicted levels.

Table III.1. **Functions of noise prediction and measurement**

	Prediction (mathematical models)	**Actions**	**Measurements**
NEW ROAD	Noise prediction	Construction	Controls of predictions (checks)
EXISTING ROAD	Prediction for enlargements of road	Enlargement	Controls of predictions (checks)
	Prediction for protections	Noise protections	Control of protections

Prediction methods have proved to be very useful and some of them have been applied in a wide range of noise situations. However, all existing prediction methods are limited by the small number of different scenarios available (only certain types of road structures are considered in the models). There are also limitations within a given scenario resulting from the reliability thresholds that are part of the simulation model (for example, some models impose vehicle speed thresholds).

Thus, the range of validity (types of simulation scenarios, reliability thresholds of each variable) of each prediction method must be taken into account in order to ensure proper assessment.

As vehicle noise emission is reduced through policies of noise emission control in most OECD Member countries, prediction methods must be continuously tested and modified to adapt the theoretical emission levels to the real changing ones.

Despite the technical benefits of prediction methods and the general trend in their favour, measurement can hardly be replaced by calculation methods when noise assessment is required in existing situations such as those involving legal procedures (for example, community appeals to government concerning noise pollution effects).

III.2. NOISE PREDICTION METHODS

Noise prediction methods have been developed in various countries for the assessment of traffic noise levels. In some countries, certain prediction methods are officially promoted or adopted by the public authorities responsible for land use planning and noise abatement design. In these cases, regulations establish the calculation algorithms used in the method and the type of result to be provided.

Generally, in countries where official methods exist, the use of another prediction model is permitted, only if the user demonstrates that the results obtained are similar to those provided by the official method.

The various noise prediction methods, characterised by different levels of detail and reliability, can be classified in three basic groups:

- **Manual methods based on nomographs, tables, or simple analytical equations**
  These methods are used for a preliminary assessment, and are applied to simple situations. A general calculation formula is used to determine the traffic noise level, while nomographs and tables are used to correct for various topographical or other conditions. Most of these methods are simplified versions of more complex mathematical models.

- **Physical scale models (see figure III.1)**
  Simulations using physical scale models allow a highly detailed reproduction of very complex spatial situations. However, they are extremely costly in terms of both money and other resources, since they involve the construction of an "ad-hoc" physical model and require highly sophisticated experimental equipment.

- **Numerical simulations by automatic calculation**
  Computer programs can be used to generate predictions for most topographical scenarios. They can evaluate acoustic propagation, reflection, and absorption phenomena. The detail and accuracy of their results depend on the complexity of the model as well as the quality of the input data.

Except when a physical scale model is used, which is rare, the traffic noise prediction is made using mathematical formulae. These formulae result from both theoretical considerations of sound propagation and empirical considerations involving emission power and certain attenuation values. The complexity of the calculation processes, especially if some degree of accuracy is desired, requires the use of computer programs to shorten the calculation time.

All models use parameters representing the different variables involved. In all cases parameters reproduce the sound sources (traffic parameters), topographical conditions (including those of the roadway), locations of reception points, attenuation by air and ground, and the presence of obstacles between the source and the receiver. Meteorological influences are not considered in most existing prediction models.

Figure III.1. **Physical scale model**

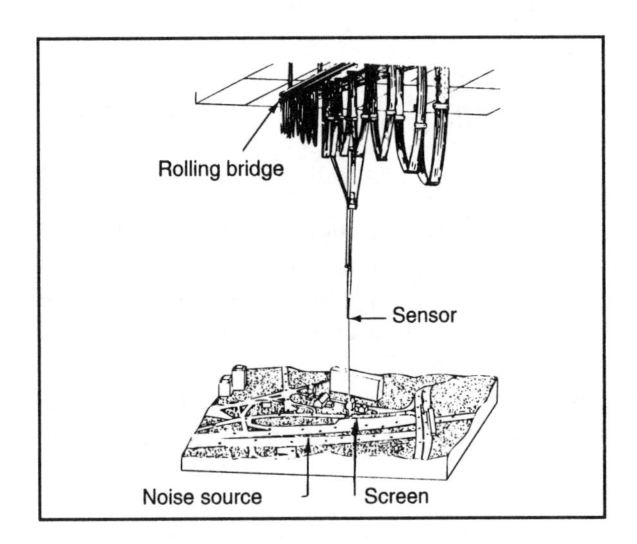

The general structure of the mathematical models is as follows:

1. Topographical description of the site, including the locations of reception points, sound absorption characteristics of the ground, presence of natural and artificial barriers, etc.

2. Definition of emission sources (roadway, railway, etc.), determining profiles, cross sections and structures (sunken, level or raised roadbed, tunnels, viaducts, etc.)

3. Acoustic characterisation of sources (traffic flow, average speed, types of vehicles, etc.)

4. Analysis of sound diffusion in propagation. The attenuation due to distance, ground absorption, reflection and diffraction by obstacles, and sound absorption by the air all have to be considered.

5. Readout of results

III.2.1. Mathematical models of traffic noise

The calculation formulae used by most prediction models are very similar. Basically, a reference noise level, corresponding to the noise level due to a single vehicle running under standard conditions at a reference distance, is obtained experimentally and incorporated into the formula as a constant. Correction factors are used to allow for the influence of the types of vehicles, traffic flow, average speed, distance, type of pavement, ground absorption, road cross section, screening effect of obstacles, etc. The number and values of these factors vary from one model to another.

The State Road Authorities in **Australia** use the CoRTN (Department of Transport Welsh Office Calculation of Road Traffic Noise, 1988) prediction method to assess traffic noise. This prediction method has been adapted for use in various computer software packages. It is used at the planning/design stage in the development of new roads and to determine noise barrier features. The

Figure III.2. **Flow chart of prediction models**

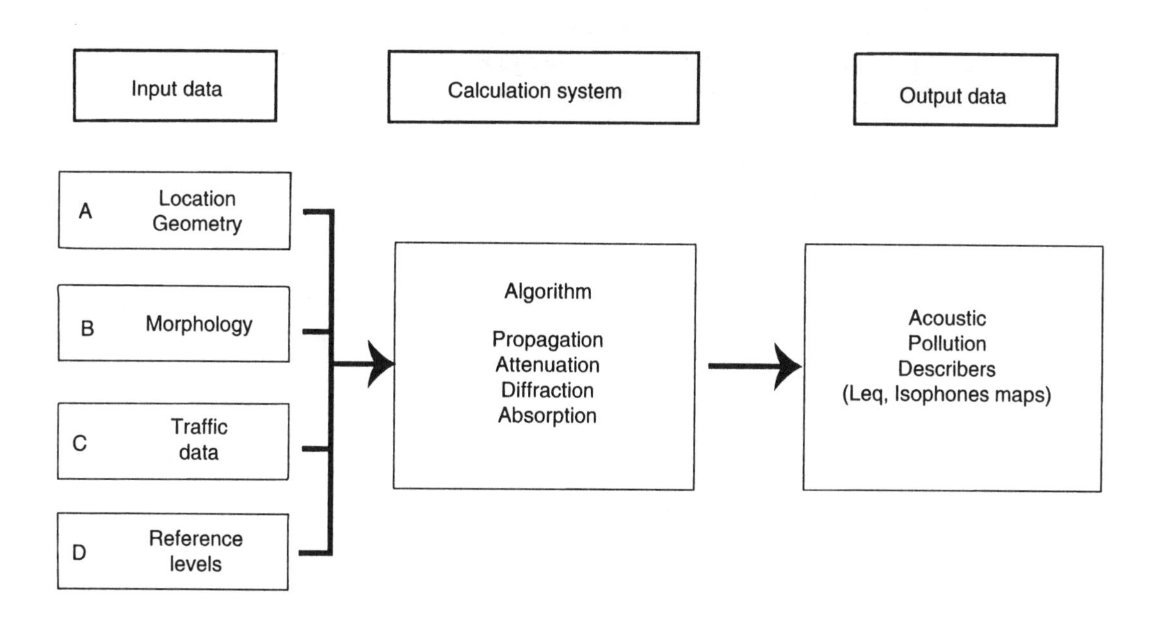

appropriateness of the CoRTN model is continuously checked by comparison of simultaneous measurements and predictions.

There is another model developed at the Australian Road Research Board (ARRB) that involves the prediction of noise levels at intersections with traffic signals. However, its scope is very limited.

In **Austria,** noise emissions are assessed using an Austrian calculating method, developed in 1983, which has official status for federal roads[1]. The hourly equivalent sound level is predicted for day time and night time according to the following formula:

$$L_{eq} = L_g + 10\log MSV_L + K_R + K_S + K_F + K_L + K_G + K_K - K_E - K_W - K_H$$

where:

L_{eq} A-valued energy equivalent sound level in dB(A)
L_G 32 dB(A) (basic value for sections without any construction and free sound spreading at a distance of 25 metres from reference line)
MSV_L Standard hourly traffic rate in vehicles/h
K_R Correction for influence of reflections
K_S Correction for the share of heavy traffic
K_F Correction for the type of pavement
K_L Correction for the longitudinal inclination of the road section

1 A new calculation method with an adequate computer software was developed within the past years and will be introduced at the end of 1995.

K_G	Correction for the standard speed
K_K	Correction for the influence of crossings
K_E	Correction for the distance from the road centreline
K_W	Correction for the length of effective road section
K_H	Correction for screening effect of obstacles

In Austria, a calculation can be based on another model only if a comparison between its results and those of the Austrian method is performed for at least one important emission point within the area studied.

A Nordic road traffic noise calculation model has been officially adopted by the public authorities of **Denmark, Finland, Norway and Sweden.** It is used in land use planning, traffic planning and design, and noise abatement design.

The model has the following structure:

$$L_{Aeq} = L_{Aeq10m} + LV + LN + LTF + LAV + LM + LK + LF$$

where:

L_{Aeq}	Equivalent noise level
L_{Aeq10m}	Reference mean emission level at a distance of 10 metres from the road centreline
L(I)	Correction factors

Correction factors are used to adjust for speed, traffic flow, heavy vehicles, distance, terrain and shielding attenuation, facade insulation angles of incidence smaller than 180°, wide barriers, road inclination, short distances, single and multiple reflections, and attenuation by low-density housing.

The model includes a comparable formula for determining maximum noise level, LAmax, which is independent of vehicle speed and traffic flow.

A computer program, NBSTÖY, based on this model, was adopted in 1990 by all Nordic countries. NBSTÖY assesses only one point at a time. The Norwegian model differs from the other Nordic methods in one point: 3 dB(A) are added to facade noise levels in the results of the calculation when calculating the noise level near a facade and will be compared to the standard. In the other countries, free-field results are compared to the standard. A new model, TSTÖY, now in development (on AUTOCAD), is aimed at assessing several points by adapting the previous model to a computer-based ground model in which ground data are digitised.

Generally, the NBSTÖY gives results in good agreement with measured values, but in some cases it overestimates noise levels. It is not very well suited to studies of different technical noise abatement measures, such as road resurfacing, limiting urban speeds to less than 50 km/h, or using special noise barrier structures. The maximum level calculation has been criticised as not representative of actual maximum levels. A revision of the model is expected for 1995.

Norway has developed a PC model (VSTÖY), including a data base, based on a simplification of NBSTÖY, for a rough, general evaluation in which the ground is particularly simplified. This model can be used to build up an environmental data base for noise.

In **Germany,** a document entitled "Directives for Anti-noise Protections along Roads" (R.L.S.-90), published by the Road Construction Section of the Federal Ministry for Transport, provides a method for predicting noise levels generated by road traffic and a method for designing anti-noise barriers. The R.L.S. method allows calculation of weighted Equivalent Level "A" as a function of road traffic data and the morphology of the area being studied. It provides two distinct calculation methods, named "A" and "B". Procedure "A" applies only to long, straight road conditions, where the effects of obstacles or traffic deviations are negligible. Procedure "B" applies to all other cases.

The R.L.S. method is an integrated calculation procedure, easily implemented on a computer. The calculated levels are generally higher than those obtained experimentally, and the method is used as a cautionary procedure for prediction of environmental impact and the designing of anti-noise barriers.

Italy does not have official traffic noise prediction models. Various methods have been applied in some studies and projects. Official methods of other countries have been practiced in most cases, although several methods, developed in Italy by various research teams, are also used.

The Autostrade company introduced the "Modello Inquinamento Rumore Autostrade" (M.I.R.A.) in 1990 in order to have an adequate prediction model for noise in areas adjacent to such infrastructure as toll highways and urban bypasses. It is a semi-empirical model based on an American development and is in two main parts: i) determination of the reference level and subsequent corrections according to traffic flow, and ii) determination of the acoustical diffusion to the receptor. It has been adapted to include emissions generated by the specific Italian vehicle composition in traffic flows and to include specific Italian road surface types. Calculation is done separately for three classes of vehicles (light, medium and heavy) and the final results are obtained by summing the levels due to each class. The main calculation formula is:

$$(L_{eq})_i = (L_0)E_i + 10\log\left[(N_i \pi d_0) / (T \cdot V_i)\right] + \text{Corrections}$$

where:

$(L_{eq})_i$	Hourly equivalent level for each vehicle class
$(L_0)E_i$	Median reference sound level for each vehicle class
N_i	Number of vehicles of the i-th class in an hour
d_0	Reference distance at which $(L_0)E$ is determined
T	Reference time (1 hour)
V_i	Average speed for the i-th class of vehicles

Corrections are made to take account of distance between road and observer, absorption by the ground, reflection from obstacles, contribution of different segments of the road, and shielding effects.

The IPSE calculation method, also used in Italy, is a traffic noise prediction model derived from the Australian "Environment Noise Model" (E.N.M.). This method uses ray-tracing techniques and a data bank that describes the Italian traffic and road condition situation. The model takes account of ground and wind effects, diffraction on obstacles, and the sound absorption properties of surfaces. The model can be connected to AUTOCAD software.

Another prediction method, COSA & NICOLI's, is used for urban traffic situations. The calculation element in this method is the SEL (Single Event Level), and the Leq is calculated according to the SEL values of 5 vehicle categories.

Evaluation of traffic noise in **Japan** is by assessment of the median sound pressure level, L50. The calculation is based on a prediction formula and can be applied only when vehicles are travelling at almost constant speeds on roads that must be continuously flat, embanked, cut out of hillsides, or elevated. Simulations and other methods for estimating interchanges, excavated roads, tunnel portals, etc., are currently being tested or developed. The prediction formula used by the method is:

$$L_{50} = L_w - 8 - 20\log l + 10\log(\pi\ l/d \cdot \tanh 2\pi l/d) + \alpha_d + \alpha_i$$

where:

L_w =	$86 + 0.2V + 10\log(a_1 + 5a_2)$
L_{50}	Median value of sound pressure level at the point of assessment
L_w	Average power level generated by one vehicle
V	Average speed (km/h)
N	Average traffic volume (vehicles per hour)
d	Average spacing d = 1000V/N
a_1	Small sized vehicle ratio
a_2	Large sized vehicle ratio
l	Minimum distance from sound source to the assessing point
α_d	Value corrected for diffraction damping
α_i	Value corrected for miscellaneous factors

In **the Netherlands,** the "Calculation and Measurements Regulations for Traffic Noise", issued by the Minister of Housing, Physical Planning and the Environment, specify two calculation methods: Standard Calculation Method 1 (SCM1) and Standard Calculation Method 2 (SCM2). This document states that it is preferable to use calculation methods in new situations, when measurement are not possible or in situations where the measurements are greatly influenced by meteorological conditions.

SCM1 is only used to determine noise levels in a specified range of application. In general terms, SCM1 can be used in situations where there is no insulation and the road does not display major variations in aspect and traffic data, i.e. in relatively simple cases with few observation points. It is often used in the preparation of plans when it is desired to obtain a quick impression of the sound pressure impact at a particular place.

SCM2 is a more sophisticated method, used in more extensive surveys and in surveys where the effect of insulation needs to be calculated. The equivalent sound level is obtained by summing the energy values obtained for each octave band.

For further details concerning SCM1, SCM2 and measuring methods, please refer to references (12) and (13).

In **Spain**, there is no generally applied formula or mathematical model in traffic noise prediction. The methodology used in each study must be specified and described.

The mathematical formulae used to predict sound levels generated by traffic on newly-built town roads and dual carriageways are similar to the formula proposed in the French "Guide du bruit", i.e.:

$$Leq = 20 + 10\log(Q_{VL} + EQ_{VP}) + 20\log V - 12\log(d + \frac{lc}{3}) + 10\log\frac{\theta}{180^\circ}$$

where:

Leq	Sound level
Q_{VL}, QVP:	Number of light and heavy vehicles

E	Factor of acoustic equivalence between light vehicles (<3.5 T) and heavy vehicles (> 3.5T)
v	Speed in km/h
d	Distance to the roadside, in metres
lc	Width of the roadway in metres
Θ	Angle at which the road is seen, in degrees

The value of equivalence factor E depends on the type of road and the gradient in percent, r.

Table III.2. **Equivalence factor E**

% gradient	**r ≤ 2%**	**r = 3%**	**r = 4%**	**r = 5%**	**r ≥ 6%**
Dual Carriageway	E=4	5	5	6	6
Highway	E=7	7	10	11	12

The speed is the median speed of all traffic (speed reached by 50% of vehicles during the period of observation). For a dual carriageway with a very wide central reservation, the study must be broken down, the final result being obtained as the total of the partial sums.

Concerning computerised mathematical models, the Ministry of Public Works, Transport and Environment has since 1992 been using the French MICROBRUIT (CETUR) and MITHRA (C.S.T.B.) models, although there is no official calculation method.

In **Switzerland,** the Federal Material Testing and Research Laboratory (EMPA, Dübendorf) have developed a traffic noise assessment programme, StL-86, at the request of the Federal Environmental Protection Office. The computer model consists of a topographical model and an acoustic model. The former uses polygons as information vectors. These polygons contain data on road and obstacle location and topography.

The Federal Highway Administration (FHWA) of the **United States** has developed and circulated a traffic noise prediction model, used by all State Highway Agencies in the United States, the FHWA Highway Traffic Noise Prediction Model. However, other traffic noise prediction methods may be used provided that the methodology is consistent with that of the FHWA model. The calculation is made separately for three classes of vehicles (light vehicles, medium trucks, and heavy trucks). The final level is obtained by summing the levels of the various classes. Attenuation due to temperature gradients, winds, and atmospheric absorption is ignored in the FHWA model. This model does not currently account for multiple reflections of sound waves.

There is computer software based on the FHWA model: STAMINA 2.0 is the traffic noise prediction programme and OPTIMA is a companion programme for noise barrier design. The resulting software output indicates the sound level at each receiver and the noise contribution from each road segment. Graphically, it depicts plan and profile views. Links have been made to CADD systems.

In the **United Kingdom,** the C.R.T.N. method (Calculation of Road Traffic Noise), developed in 1975, allows the evaluation and prediction of statistical level L10, on an hourly basis or for the period between 6 a.m. and 12 p.m. The calculation procedure, subdivided into numerous equations and charts, is applicable for distances from the road shorter than 300 metres and for wind speeds below 2 m/s.

The L_{10} level can be converted to the equivalent level LA_{eq} by one of the methods suggested by various authors. In most cases it is sufficient to subtract 3 dB(A) from the L_{10} level to obtain the corresponding LAeq value, with a margin of error that is within ±2 dB(A) in 95 per cent of cases. The method was recently updated to extend its scope to smaller vehicle flows and achieve more reliable evaluation in unusual configurations, such as sunken roadbeds and the presence of acoustical barriers on both sides of a roadway.

Figure III.3. **Noise reduction with a barrier due to a longer noise path**

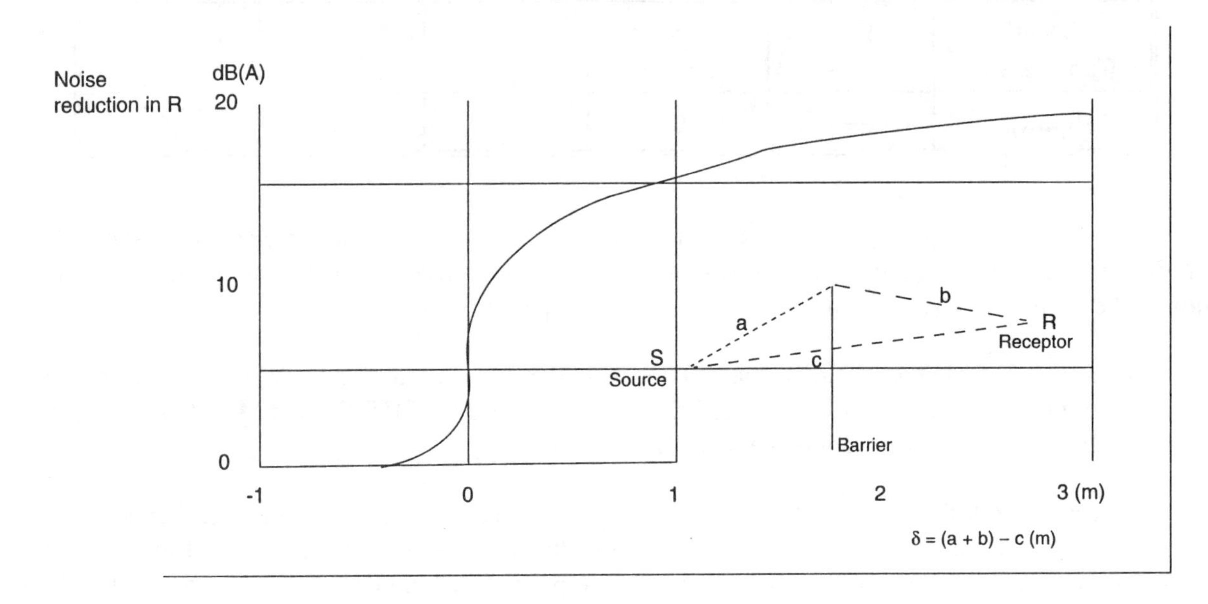

δ : additional noise path due to the presence of the barrier

The method calls for the determination of the L10 level at a reference point 10 metres from the edge of the roadway. The reference level is later corrected for road slope and type of pavement. Further corrections are made for major factors influencing sound propagation, such as sound attenuation by distance, sound absorption by the ground, the visual angle between source and receiver. The effect produced by natural or artificial barriers can be evaluated using the chart of figure III.3.

On behalf of the **Commission of the European Communities,** a group of researchers at the "Laboratorium voor Akoestic en Warntegeleiding" of the Catholic University of Leuven (Belgium) has prepared a document entitled "Guideline for the calculation of traffic noise". The final report describes methods used to calculate traffic noise levels both in urban areas and in open countryside. Estimated sound levels can be expressed in LA_{eq} and in L_{10} dB(A), and are obtained using charts and multi-entry nomographs.

The results obtained using this method have been evaluated as follows:

- For close positions, from 2 to 4 metres above the ground, the procedure generally gives results higher than those obtained experimentally.

- For heights above 5 metres, the results obtained are lower than the experimental ones, probably because, with the receiver at this height, the assumption of reduction by visual angle is not verified.

III.2.2. Analysis of calculation models

A traffic noise prediction method must be capable of determining noise levels under any conditions of emission and propagation. Moreover, it should provide results accurately representing real existing noise levels. Achievement of these aims depends on the following two assessment processes:

- assessment of noise emission levels due to traffic flow
- assessment of noise attenuation between the source and the reception point.

Calculation of traffic flow noise emission levels is generally based on noise levels produced by different categories of vehicles, at a predetermined distance, under free-field propagation conditions. Levels at the reception points are calculated using acoustic propagation laws to estimate attenuation between source and receiver.

Emission levels are the input data for the methods. They have been obtained by noise measurements carried out in various countries for their specific types of vehicles. However, there are many variables in addition to the number and type of vehicles that determine the global emission level (type of pavement, dynamic traffic flow characteristics, gradient of the road, etc.). Furthermore, even if these conditions are taken into account, the calculated emission levels may be erroneous because the reference levels do not consider the age and maintenance of the vehicles.

Work is needed to create data bases, which would be permanently reviewed and updated and contain vehicle noise emission reference levels for all vehicle categories, types of pavements, and traffic conditions, to allow the users of prediction methods to select the correct emission input data in each situation. Moreover, improvements concerning the spatial distribution of noise sources within the road are needed (e.g., various lanes with different types and numbers of vehicles as opposed to a single lane with mixed traffic).

Concerning the attenuation of sound energy in propagation, limitations on the reliability of methods mainly affect the assessment of sound energy losses due to the ground effect and by the presence of obstacles. The prediction methods generally calculate sound levels higher than those obtained experimentally, since they use a highly simplified view of all the complex attenuation phenomena. Measured levels for reception points close to the source are usually lower than predicted because the barrier effect of the vehicles passing on the multiple lanes is not considered. Significant differences have been also found between predicted and measured levels at distant reception points, where sound attenuation caused by the ground effect becomes large. The presence of obstacles and complex topographical conditions often produces important deviations from calculated levels. These effects are the greatest weakness of most prediction methods.

The most widely used computer programs for traffic noise prediction offer a large range of calculation possibilities. In a relatively short time, predictions can be made for a large number of reception points, with different traffic and noise barrier assumptions. However, despite the general development of computer techniques to incorporate spatial variables, existing prediction programmes have not managed to completely assimilate topographical obstacles, unless the obstacles are handled by the user as standard barrier types. This factor definitively undermines the application of these methods to environmental impact assessments of extensive areas where topographical conditions are not as legible as the accuracy of prediction methods would require.

Mathematical models for noise assessment in urban areas (BURGESS, GRIFFITHS, LANGDON, COSA & NICOLOSI, etc.), have been developed according to acoustical theories of sound propagation in semi-reverberant fields. These models often give results of low accuracy, because sound emission and propagation conditions in urban areas are not easily simulated. Some of these methods have been applied with success in particular conditions, but they must be used with care.

III.3. NOISE MEASUREMENT METHODS

As was mentioned above, in traffic noise assessment, calculation methods are generally preferred to measurement methods for technical reasons. The measurement methods used to determine the specific acoustic characteristics of the materials used in anti-noise barriers or facades are regulated by national and international standards organisations. The sound-insulating and sound-absorbing capacities of materials are calculated from measurements made in anechoic chambers (mainly ISO-140, ISO-717 and ISO-354), since open-field measurements (French impulsive method and others) have not demonstrated good performance or accurate results. The methods described in this chapter concern only open-field measurements, the main goal of which is to measure, at a certain number of points in a given area, the noise levels due to road traffic. Some further discussions on measurements and the assessment of the effectiveness of low noise surfacings and noise barriers is to be found in chapters V and VI.

III.3.1. Methodologies

Measurements are made mainly for the following reasons:

- to determine noise levels in an area so as to identify undesired situations,
- to compare background variations in noise levels,
- to compare noise levels before and after road construction, and
- to estimate the effectiveness of the anti-noise measures implemented.

Locations, measurement times, and the methods selected depend on purpose and scope. For example, when a large area must be evaluated, the selection of measurement durations and survey points greatly affects the significance of the results obtained.

Traffic variation analysis and a survey of the territorial distribution of activities should be conducted to select the points where measurements will be made. The aim of these analyses is to optimise the time spent in measurements while keeping the results fairly accurate and representative.

Noise measuring is a hard process that requires time and specialised instruments. In addition, some precautions must be taken to guarantee the precision and reliability of the measurements. These precautions are as follows:

- using proper testing procedures and methods,
- checking instruments regularly (recommended once a year),
- calibrating instruments before and after use,
- making sure that instruments are not affected by weather conditions,
- recording weather conditions during testing (wind speed, humidity, etc.) with a view to accepting or rejecting the results,
- avoiding making measurements under exceptional conditions (rain, snow, or ice).

III.3.2. Measuring instruments

The Leq, in dB(A), is the most important and frequent evaluation parameter used in road traffic noise assessment. Measurement equipment that allows the direct acquisition of this value, known as integrating sound level meters, is preferred. Since road traffic noise varies with time, the most useful instruments are those designed for continuous measurement of Leq. Actually, there is a broad range of acoustical instruments, portable or not, designed for short and long measurements, that yield many different noise indexes (Leq, MaxL, MinL, LN, SEL, histograms and others). Some are even designed to help in the data processing. Equipment should be selected according to measurements goals, and it is important to be sure that the instruments are valid for outdoor use.

Sound Level Meters are classified into different types or classes according to their precision. The IEC 651 (International Electrotechnical Commission Standard 651) classification is generally used as performance requirement reference for sound level meters. Depending on the measurement aims, different classes are recommended or required. Class 2 or better instruments are required for general measurements, while Class 1 may be necessary for detailed assessments.

III.3.3. Time and intervals

Noise levels due to road traffic vary spatially and in time. Statistical sampling techniques must be used for the accurate determination of the acoustical environment of an area. A distinction must be made between measurements performed for rough assessment and those for detailed evaluation of specific features. The latter are necessary when precise noise levels must be established for specific situations; when the effectiveness of anti-noise actions is to be assessed; and when measurements are to be made in significant or reference points.

Different durations are chosen according to the intended goals and traffic features. There are different approaches to the choice of measurement durations:

- Determine the hours when traffic is greatest and measure a mean value for that period.

- Measure for the time corresponding to the passage of at least a certain number of light and/or heavy vehicles, and consider the results obtained as the typical sound energy of the road.

- Measure over long periods (more than 24 hours).

As a general rule, measurement periods should be as long as necessary to provide a good knowledge of day-night, weekly, and seasonal noise variations for the area's average weather conditions. Hence, there is no fixed maximum measurement time. However, as studies and assessments must often be made in short periods of time, most national regulations have established minimum measurement times according to assessment goals.

The Nordic NORDTEST measurement method (Denmark, Norway, Finland, and Sweden) recommends a minimum duration of 15 minutes or the passage of 500 light vehicles or 50 heavy vehicles for daytime traffic flow. Night levels are usually calculated from daytime measurements on the basis of traffic flow data. In the United States, 15-minute periods in the noisiest hour are typical; nevertheless, measurements up to 24 hours are made if there is any public controversy at a specific location or if information to identify the noisiest hour is not available. In Austria, the measurement time depends on the traffic distribution; 30 minutes is typical. In the Netherlands, the minimum duration is 10 minutes or 100 light vehicles and 10 other vehicles. In Japan, at least one measurement must be taken for typical noise conditions during each of the following periods: morning, daytime, evening, and night.

However, when the traffic flow distribution is not available, measurements up to 24 hours should be made for proper evaluation of daylight and night noise levels. If traffic data are known, shorter intervals can be selected, but an interval of 15 minutes or 500 passing vehicles is the minimum acceptable measurement period. Noise assessments of roads with large seasonal traffic flow variations must take high-season noise levels into account.

III.3.4. Survey points

The number and locations of survey points needed to identify the sound environment of an area depend on the type of measurements to be made. The selection criteria can be summed up by two general criteria:

- select points where people are likely to be disturbed by road noise, and
- select points which are representative of different situations and conditions in the area.

The first criterion is intended mainly to identify black spots and ensure the quantification of noise levels in areas where roads have been built and are in service. In these cases, the points selected must represent the exposure conditions of the largest possible number of the people exposed to road noise. Significant points are not necessarily those locations where noise levels are the highest. Typically, significant points are on the outsides of the buildings that are closest to the road, sampled at different heights. Measurements must be made outside the buildings at a distance of 1 to 2 metres from those facades closest to the road. If the points are in front of or very close to windows, it must be specified whether the measurements are made with the windows open or closed, so that the reflection effect can be assessed. Figure III.4 shows some possible locations for measurement points.

When human activities are carried out in open areas (gardens, strolling areas, walking paths, sports fields, etc.), measurement points may be located where the highest noise levels are expected. The microphone must be at least 1.5 metre above the ground.

Figure III.4. **Location of measurement points near facades**

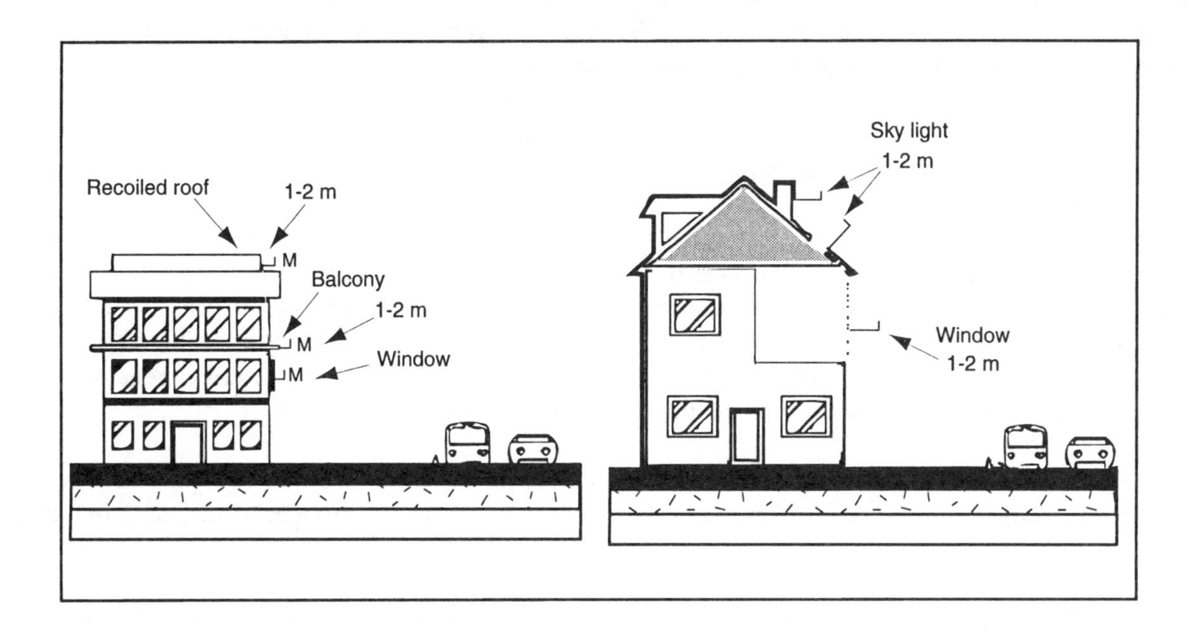

The second criterion is applied to both existing and pre-operational situations and concerns the spatial distribution of noise levels in an area, for example, in plotting acoustic maps or checking prediction results. The selection of representative points involves defining "homogeneous sectors" that include roads and reception points. In practice, the procedure is to divide the area into sectors where sound emission can reasonably be considered constant and topographical characteristics do not vary much. Measurement points are established for each sector so that noise levels at any point within the homogeneous sector can be deduced from the measurement results.

Nevertheless, it is very difficult to design universal methodologies for outdoor noise measurements. Different situations require different methodologies. It is generally agreed that the use of minimum durations designed for well-known standard situations may not be enough for an adequate assessment in other situations, which are unfortunately common in practice. The selection of survey points depends on the scope of the study and on noise emission and propagation conditions within the area.

It is common practice to correct measurement results to calculate standard daytime and nighttime levels. The methods provide formulae for compensating differences in traffic conditions and in Leq reference time. These corrections depend on the availability of traffic flow data.

III.3.5. Measuring methods inside buildings

The sound level inside buildings is determined to establish outside wall insulation needs and to check whether the legal interior requirements are being met. Measurements must be made both inside and outside (next to the window) at the same time. This calls for the use of a two-channel analyser or two sound level measuring instruments.

Some countries (Norway, the Netherlands, etc.) have developed calculation methods for assessing indoor noise levels. These are based on calculations of noise reduction from data on ventilation, room volumes, acoustic conditions of windows, etc. The insulation values of the separate facade components are taken from data files accessible to computer programs.

III.3.6. Measuring methods for noise barrier effectiveness

An anti-noise barrier must be acoustically tight. The properties of materials or panels used in barriers can be tested in anechoic chambers determining sound insulation in accordance with ISO 140 and ISO 717 and sound absorbance in accordance with ISO 54. Nevertheless, the effectiveness of a barrier depends on its final properties, once built. Construction elements and procedures (junctions between structures, expansion joints, support elements, etc.) also determine the effectiveness of the barrier, in addition to the acoustical properties of the materials or panels.

Noise barrier effectiveness is determined by measuring the field insertion loss provided by the barrier. In some countries, the effectiveness of barriers is controlled by special test of "simulated insertion loss".

The field insertion loss is the noise level at a reception location before a barrier is built minus the noise level at the same location after the barrier is built. If noise barriers could be built instantaneously, the insertion loss could be easily determined and would be the difference between the "before" and "after" measurements. However, because of the time required to build a barrier, new factors are introduced, for which corrections must be made. These factors include changes in traffic volumes, mixes, and speeds, changes in emission levels, and changes of the ground. Barrier insertion loss measurements must take these factors into account; reference microphone locations are used for this. Depending on site circumstances, reference microphone locations can be in front of the barrier, above the top of the barrier, or beyond the end of the barrier. To determine the barrier insertion loss along an existing road where the barrier has not yet been built, simultaneous measurements are made at the reference and study site microphone locations. Two sets of measurements are made: one set before the barrier is built and one set after. If necessary (because of changed conditions), an adjustment is made to the reference measurements and the insertion loss is calculated by subtracting the "after" measurement from the adjusted "before" measurement.

The procedure for determining the insertion loss for a barrier along a new road or along an existing road where the barrier has already been built combines "after" measurements and the results provided by a calculation method. A set of measurements is made and used to calibrate the calculation model. Reference locations are again used in the measurements. Once the calibration is completed, a "before" noise level is calculated. The insertion loss is then determined by taking the difference between the calculated "before" and the measured "after" noise levels.

A report regarding the test measurements must be written, including information about barrier geometry and material characteristics, the site where measurements are made, traffic flow, and the acoustic instruments used for the test.

Tests of simulated insertion loss are conducted on a barrier of standard dimensions on a flat highly reflective zone (e.g., smooth cement, non-porous asphalt, or similar), without reflecting obstacles around the test site. In all cases, the type of ground must be precisely described in the test report. Wind speed during measurements must not exceed 2 m/sec.

Figure III.5. **Example of open field measurement**

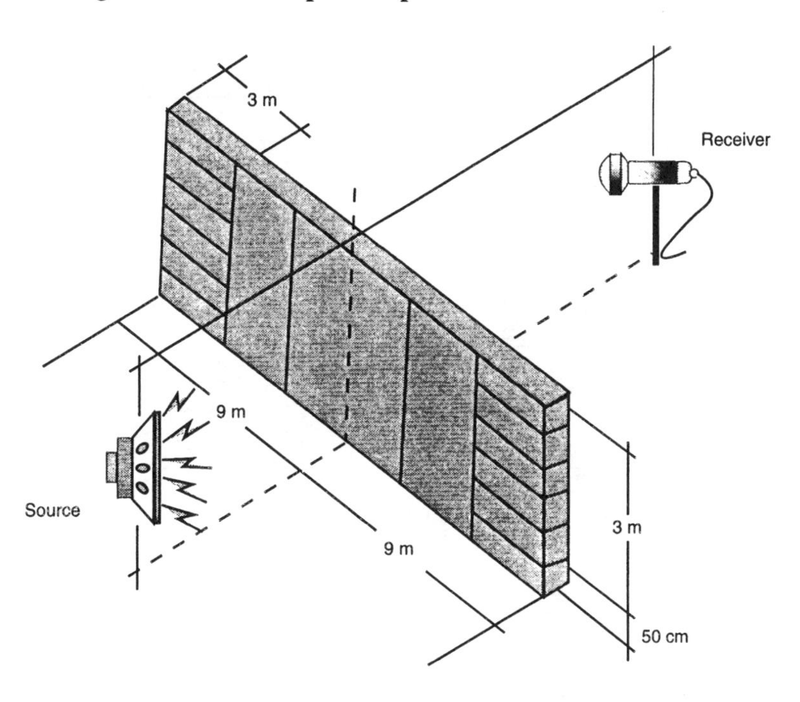

A loudspeaker 0.15 m or less in diameter is used as sound source. Its directional characteristics must be measured on site and be reported in the test certificate.

The barrier will have standard dimensions (18 m long, 3 m high are suggested) and its lower edge will rest on a bed of sand to seal off the foundation zone; any uprights or other barrier structural elements must be placed in the same positions as in normal use.

Two different tests are performed:

- *Measurement of insertion loss:* this assesses the attenuation provided by the barrier under standard conditions, possibly revealing effects of special devices modifying attenuation by diffraction (T-shaped crest, waveguides, etc.). For this test, the sound source is located 3 metres from the barrier and 1 metre high. The receiver is placed 25 metres from the barrier and by turns at 1 metre and 2.5 metres high. The source-receiver axis must pass through the mid-point of the barrier. Measurements must be made without and with the barrier, yielding the insertion loss values as the difference between the two measurements.

- *Measurement of sound insulation:* the aim of this test is to assess the acoustic insulation provided by the barrier under standard conditions, revealing any negative effects of joints, acoustic bridges, acoustic holes, etc. This test cannot replace assessment of soundproofing capacity (ISO 140 and ISO 717). For this test, the source and receiver are both located at 1 metre high and 1 metre from the barrier, on opposite sides. The source-receiver axis must pass through the mid-point of the barrier. Measurements are made with and without the barrier, yielding the insulation value as the difference between the two measurements.

In France, a test has been developed to calculate the sound-insulating and sound-absorbing capacity of constructed barriers. This test, called the Impulsive Method (AFNOR S 31-089), is an on-site test, and can determine the acoustic properties of a whole barrier (an anechoic chamber test determines only the acoustical properties of panels or materials).

Figure III.6. **Determination of source directivity**

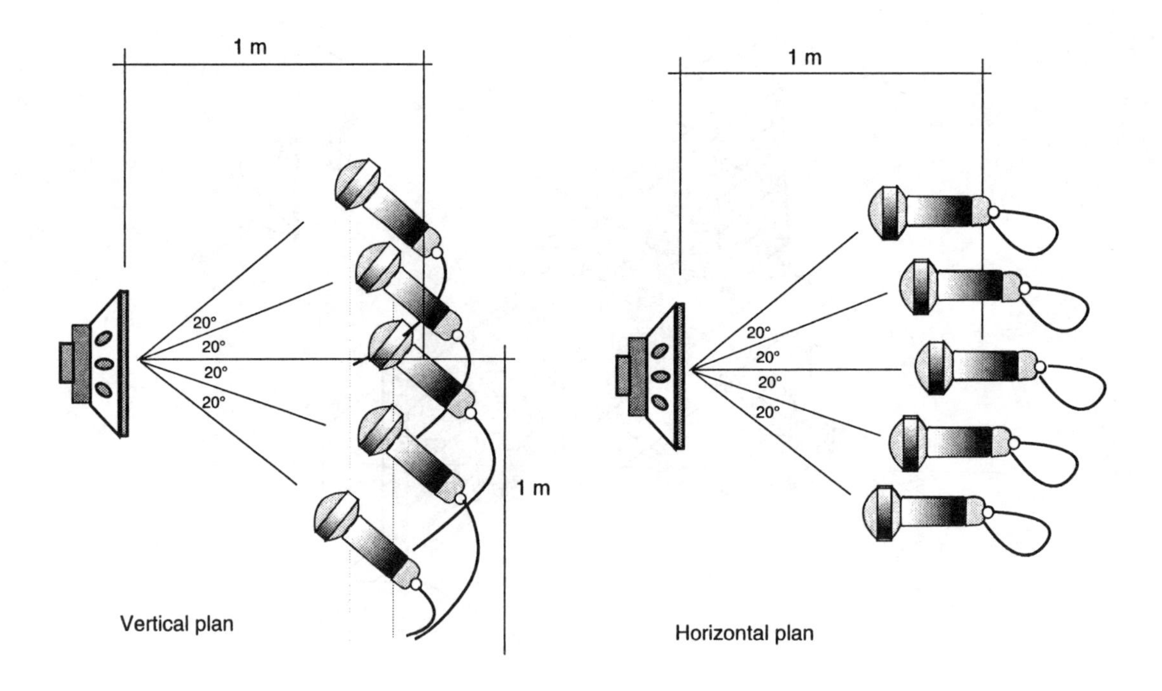

III.4. CONCLUSIONS

There is uncertainty in assessing the degree of disturbance suffered by people from road traffic noise. However, quantitative assessment of traffic noise levels is necessary for an objective analysis of noise impact in areas near road infrastructures. This assessment is based on the results of prediction and/or measurement methods. Prediction methods are preferred because they can simulate a wide range of situations and can be applied to the planning process. Measurement methods are used mainly to assess an existing situation, by taking data directly.

Prediction methods must be reliable and operative to be useful. The reliability of the results depends not only on the correctness of the mathematical formulae constituting the theoretical model structure but also on an adequate supply of input data. For a detailed noise study, the user must take a number of decisions to guarantee a proper site simulation. Traffic flows, types of pavements and ground, the acoustical properties of barriers, propagation conditions, etc., must be established by the user prior to the application of the method. Bad assessments are frequently made because input data are wrongly chosen, despite the recommendations suggested by methodological considerations and the use of decision support programmes.

Methods must cover all possible situations and calculate noise levels with a good degree of accuracy. To confirm the accuracy of results, measurements are made and compared with predictions. Prediction methods developed by OECD Member countries have more or less shown their reliability in standard situations. Nevertheless, all of them must be improved by adding better calculations of specific items. Often, viaducts, tunnels, obstacles, the ground, and other features are not adequately taken into account in calculation models.

The main calculation formulae are similar in all of the methods. Differences arise in calculating noise emission levels from moving vehicles, or in correcting for attenuation effects. All of these methods are semi-empirical, which means that theoretical formulae are combined with experimental results. In this sense, the methods are based on a large set of noise data obtained by sound measurement techniques. These data concern the emission power of vehicles, absorption characteristics of pavements, ground and materials, sound diffraction on obstacles, the influence of weather conditions, etc.

Computer models based on these methods usually do not allow modification or adjustment of the reference data used. For this reason, models designed for a specific country (which means a specific set of vehicles and traffic conditions) may not be reliable if used in other countries. It would be very interesting to allow users access to these data. Emission and other data could then be modified to adapt the calculations to the specific conditions of each country. In any case, research must be pursued continuously to update the model input data, especially in regard to the noise emission of vehicles.

In addition, computer models lack definition in topographical input data. Changes in topographic shape, ground texture, the geometric characteristics of barriers, and other features are very often neglected or incorrectly taken into account. The development of 3D programmes is expected to be very useful in improving traffic noise calculation programmes.

From a practical point of view, an interface between prediction programmes and graphic design programmes (AUTOCAD or similar) has proven to be very useful, for both data input and data output procedures. A digitised site in a common format could easily be introduced into prediction programmes with no problems of compatibility. On the other hand, a link to a graphic design programme enormously increases the possibilities for information transfer and communication.

Measurement methods are based on the direct acquisition of sound data using specific instruments. Nowadays, there is a wide range of sound level meters and systems adequate for this purpose. International and national standards define the quality of instruments according to their precision.

It is very important to emphasize that measurement is relevant only when applied to current situations. Strictly, measurement results relate only to the place and time where the measurements were made. Most measurement methods include correction formulae and parameters to adapt their results to standard conditions.

Noise measurement is a technique that can be applied in open-field areas or under controlled conditions in anechoic chambers. Measurements of the latter type, aimed at determining the acoustical characteristic of materials, are regulated by international standards. Open-field measurements are made under real conditions and are the basic measuring techniques for traffic noise assessment.

Formal measurement methods focus on the selection of time periods and measurement point locations. Different measurement durations are chosen according to the aim of the study. Trends point to noise level acquisition for periods up to 24 hours for a proper assessment of daytime and night conditions.

When variations of traffic flow with time of day are known, the durations can be shortened and the results corrected by mathematical formulae. The shortest measurement period generally used is 15 minutes. Even under these circumstances, it is recommended that long-term measurements be made at least for one point within the area.

Measurement points must be selected according to human activity in the study site. Preferred locations are those where both the highest levels are expected and human activity takes place or is likely to take place.

In determining the effectiveness of noise barriers, the calculation is done by subtraction between the measured noise levels in the "before" and "after" situations. Measurements must be made at the same time at a reference point without the protection provided by the barrier, to check changes in noise emission and propagation. If variations in conditions are detected, corrections should be made in the calculation.

Assessment methods, whether for prediction or for measurement, must be used carefully. Road traffic noise varies from one site to another and from one period of time to another. The measuring procedures and calculation methods described in national regulations can be considered as a minimum guarantee for a proper assessment. However, the validity of these techniques is not the same in all situations. Users of these methods must know their scope and comparative accuracy. They must select the appropriate methods and input data according to the goals set for the assessment.

Table III.3. **Comparison of noise level assessment methods**

Criterion	Measurement	Equations and graphs	Calculation	Computer programs	Models
Accuracy in normal use (dB(A))	± 1.5	± 3		± 2	± 2
Form of results	For a particular point in time	Outline results	For a particular point. Optimisation possible	All forms. Optimisation	Complete but no optimisation
Time taken	As necessary (1 to 7 days)	1 day	1 day*	2 weeks*	2 months
Flexibility Study of variants Re-adjustments			Fairly good	Good	Fairly good
Cost	Medium	Very low	Low	Medium	Very high
Educational value	Good	Good	Mediocre	Poor	Very good

* For a given area, noise assessment is faster with computer programs than by simple calculation; but computer programs are generally applied to larger areas, which explains why the "time taken" is 2 weeks.

III.5. REFERENCES

1. BARRY, T.M. and J.A. REAGAN (1978). *FHWA Highway Traffic Noise Prediction Model.* FHWA-RD-77-108. Washington DC.

2. CETUR (1980). *Guide du bruit des Transports Terrestres. Prévision des niveaux sonores.* CETUR, réédition 1990. Bagneux.

3. CLIFFORD R. BRAGDON. *Noise Pollution.* University of Pennsylvania Press.

4. DEPARTMENT OF TRANSPORT AND WELSH OFFICE (1988). *Calculation of road traffic noise.* HMSO. London.

5. KURZE, U (1988). *Prediction methods for road traffic noise.* Seminar on road traffic noise. Grenoble.

6. NELSON, P. (1987). *Transportation noise.* Butterworths & Co. Borough Green, Sevenoaks, Kent TN 158 PH.

7. NORWEGIAN PUBLIC ROADS ADMINISTRATION (1991). *Brukerveileder for NBSTOY 4.0.* Norwegian Public Roads Administration. Oslo.

8. NORWEGIAN PUBLIC ROADS ADMINISTRATION (1995). *Brukerveileder for VSTOY 3.8.* Norwegian Public Roads Administration. Oslo.

9. NORWEGIAN PUBLIC ROADS ADMINISTRATION (1994). *TSTOY, Terrengbasert beregning og analyse av vegtrafikkstoy.* Norwegian Public Roads Administration. Oslo.

10. TOBUTT D. and P. NELSON (1990). *A model to calculate traffic noise levels from complex highway cross-sections.* Report RR245, Transport and Road Research Laboratory. Crowthorne.

11. NATURVÅRDSVERKET (1989). *Beräkningsmodell för vägtrafikbuller.* Naturvårdsverket Förlag Distributionen, Box 1302, 171 25 Solna.

12. MINISTER OF PUBLIC HOUSING, PHYSICAL PLANNING AND ENVIRONMENT (1982). *Measurement and calculation prescription noise load inside buildings.* Staatscourant 1982, nr 228, The Netherlands.

13. DEPARTMENT OF PUBLIC HOUSING, PHYSICAL PLANNING AND ENVIRONMENT (1991). *Costs of noise reducing devices.* Report DG0 91-02. The Netherlands.

CHAPTER IV

ANTI-NOISE DESIGN AND LAYOUT OF ROADS

IV.1. THE NEED FOR NOISE CONTROL AND PRIORITY SETTING

The right measures can be implemented at the right locations, if an appropriate survey of the prevailing noise conditions has been undertaken in the different areas concerned. A number of countries (Spain, Switzerland, The Netherlands, Norway) report that this avoids new noise problems in the future, and assures the implementation of measures that reduce the number of people affected.

A noise survey provides a good basis for improving areas that are particularly affected and for deciding on which areas to place priority. It also provides a basis for judging the consequences of different developments and control measures. Having a data base of the persons affected by noise, it is also possible to predict the noise situation and evaluate the effect of measures. Noise data bases are being built up in several countries. These can, for example, be linked to traffic distribution models and/or sets of methods to provide information about the effects of different measures.

The criteria by which different countries judge the need for noise improvement vary. They all agree however that the noise factor must be evaluated when new roads are being built. Noise problems must also be avoided in the case of new buildings. However it appears that this does not have as high a priority as in the case of new roads. Several countries have, moreover, realised that noise problems need to be addressed and can be solved not only when building new roads, but also by introducing noise control measures along existing roads. The existing road network is usually comprehensive, and new roads constitute only a small part of the network. Improvements must be made along the existing roads, if a significant reduction in the number of persons exposed to noise is to be achieved. Some countries (such as The Netherlands, Switzerland, Austria, Norway, Japan) invest in special 'noise clearance' programmes for improvements along existing roads.

When implementing noise-reducing measures, there are other considerations that have to be taken into account. One of these is the aesthetic aspect, the importance of which varies between countries.

Practice varies with regard to which type of land use is considered the most sensitive to noise and must therefore be given the highest priority in noise abatement. Houses usually have priority. A survey of the different functions of the noise-exposed areas along roads will provide a basis for considering priorities.

International standards should not be set for the prioritisation criteria, since these are mainly based on cultural and political values.

IV.2. DIFFERENT FORMS OF NOISE CONTROL

It is necessary, first of all, to undertake preventive actions, aimed to correct urban planning and achieve appropriate organisation and regulation of transport systems. In this way it is possible to ensure greater effectiveness of constructions or other measures for noise protection. There are several ways of reducing noise or systemising the different kinds of measures. In Italy, measures provide either active or passive protection. Active noise protection consists essentially of measures to reduce noise at the source (acting upon the vehicle and the pavement). Passive protection consists of intervention designed to reduce noise during its propagation, i.e. in the course of its travel between the source and the receptor (noise barriers, artificial tunnels, etc.).

The Group has chosen to describe measures as follows:

- Physical measures for roads and/or surroundings (section IV.2.1);
- Traffic control (section IV.2.2);
- Reduction of noise at the source (section IV.2.3).

This chapter concentrates on the description of physical measures for roads and/or the surroundings, considering mostly the technical and noise-effective aspects. Chapters V and VI will describe in detail specialised anti-noise pavements and barriers.

IV.2.1. Physical measures for roads and/or surroundings

Physical noise-reducing measures for roads or their surroundings should be included in land-use planning. This is necessary in order to achieve a better overall plan and to avoid new problems. Noise abatement measures should be used in conjunction with improvements of other environmental problems that are created, or may be created, by the road or the traffic on it.

Roads and their immediate environments have different forms and functions, resulting in a wide variation in environmental quality for users (see figure IV.1). It is important that these be adapted to each other, and that this be taken into consideration when implementing various physical noise-reducing measures.

Many noise-reducing measures have been implemented with varying success in different countries. What follows is a description of these measures and some of their results.

Development of New Roads

In an area where a large number of people are subjected to noise, a new road could relieve or reduce the existing noise problem. However, a new road is seldom constructed primarily for that purpose. As a rule, other considerations weigh more heavily, such as the need for better access. A new road may mean that noise-sensitive areas will be subjected to noise, unless the acoustic situation is analysed and taken into account at the planning stages.

Figure IV.1. **Roads and streets in their surroundings**

It appears that in the last decade most countries have tried to avoid noise problems when building new roads. This means that earlier concepts are revised to take noise problems into consideration for new roads, even if this involves more extensive and more expensive solutions.

Planning the design of the road and its position in the terrain with noise attenuation in mind can produce favourable solutions from that point of view. The alignment and position of the road in relation to the terrain will give different degrees of noise reduction. A description of the results achieved so far is given below.

Road structures must take into consideration various factors including pollution, vibration, cost, safety and visual obstructions. Cost-benefit as well as other analyses of the consequences will be made prior to construction.

Adjusting road design and alignment to the terrain. Roads that follow the terrain may be favourable from a landscaping point of view (see figure IV.2.a), but will be a poor solution when it comes to noise, unless the terrain between the road and the receiver constitutes an obstable to noise propagation. Roads on embankments, in cuttings, on viaducts (figure IV.2b, c and d) or with mounds along the road will reduce the noise compared with roads following the terrain. The noise level decreases as the distance from the road increases, regardless of the location of the impact point. Figure IV.2 shows how traffic noise is propagated.

Figure IV.2. **Different noise level from roads with different alignments**

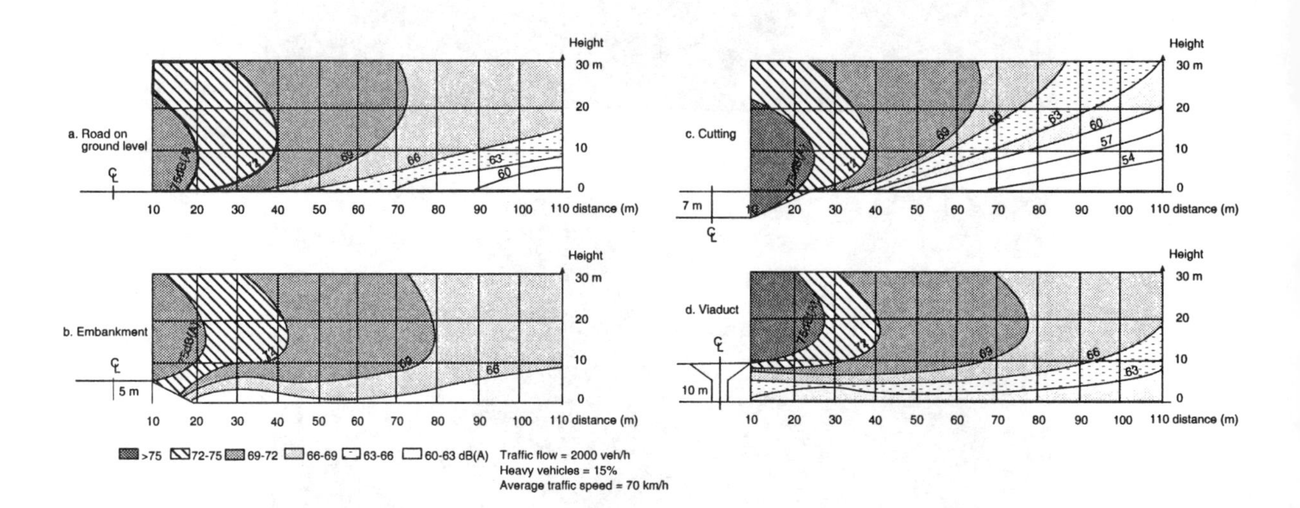

Whatever the situation, viaducts, cuttings, embankments and tunnels all have a role to play in the harmonious construction of new transport infrastructures. This is a complex matter and, besides cost considerations, certain choices are often dictated by the prevailing features of the terrain.

Roads in cuttings are usually effective in reducing noise. The noise can be reduced, compared with roads that follow the terrain, by 5 to 10 dB(A) depending on how deeply sunk the road is built (see figure IV.2.c). High noise reduction requires slopes with soft soil, preferably planted with shrubs and trees and giving protection for the neighbouring population. The slope's gradient must be as steep as possible to give the maximum protection. In urban and built-up areas this solution will be costly and require technical solutions for water drainage and because of the cost of special technical arrangements and the transport of earth. The terrain and the soil will also be of importance. Noise barriers on the top of cuttings (see Chapter VI) have proved effective in increasing the noise-reducing effect, and might also help to reduce the costs. Roads in cuttings are often a better and more cost-effective solution in rural areas.

Figure IV.3. **Acoustic performance improvement of cuttings with sound-absorbing vegetation**

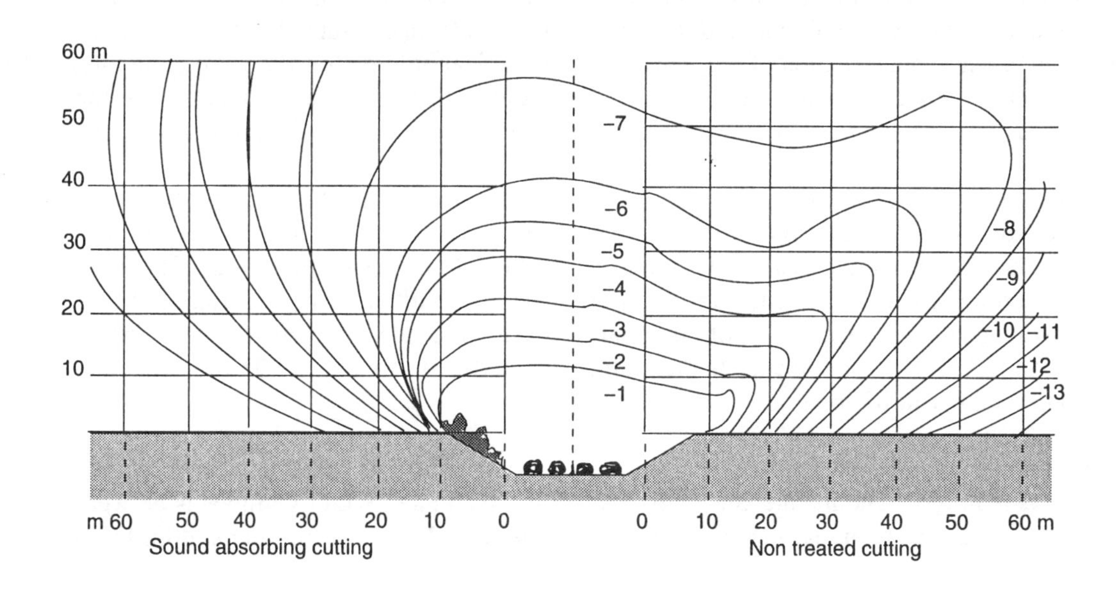

Even roads on embankments are more effective than roads that follow the terrain (see figure IV.2.b). The embankment, however, must be more than 2.5 m high. If the soil volumes necessary are considerable, this might be an expensive solution. The slopes should be as absorbent as cuttings. To reduce the noise even more, barriers on the top of the slope might be a solution, although it is not very aesthetic.

Roads in tunnels. There is no doubt that, from the point of view of noise and available area, the best solution in towns and built-up areas is to have the road in a tunnel. The ecological effect is that the environment is not disturbed by the road or the traffic. On the other hand, the solutions are costly and represent certain risks of impact both at the construction and operating stages. Tunnels can give travellers claustrophobia, the exhaust problems can cause problems inside and near the entrance. Tunnels require lighting, ventilation and drainage. The benefit of the tunnel is that land requirements are minimal and it avoids creation of the "barrier" effect typical of large transport infrastructures. There are many factors, however, which indicate that it is not always easy to build tunnels: the topography and, not least, the ground conditions, water drainage and technical regulations relating to these.

A tunnel will often be expensive and will also incur considerable maintenance costs. The experience from Norway shows that there are great variations in the cost of a tunnel. In urban areas, a two lane tunnel (in hard rocks) costs about 10 million dollars per km, a four lane tunnel about 15 million and a six lane one about 30 million dollars. In suburban areas, experience shows that the cost of a two lane tunnel is about 5 million dollars per km. The relatively low costs in Norway are the result of high technical development. In Australia estimated costs are about 130 million dollars per km for a planned tunnel in Melbourne.

The maintenance cost varies a lot and might be rather high, especially when special technical equipment is required. The maintenance cost for a four lane tunnel with two tubes is between 50 000 and 300 000 dollars per km and per year, and a six lane tunnel between 65 000 and 600 000 dollars.

With use of new technology, tunnels built in hard rock might be a better and in some cases cheaper solution compared with a road combined with necessary noise protection in built-up areas. This has achieved a considerable reduction in the number of persons subjected to noise, particularly in the largest urban areas. Good technology has been developed, making it possible to reach solutions that are satisfactory from both a financial and a practical point of view. Measures have been applied to try to solve the pollution problems by using ventilation towers and reducing the dust within the tunnel. In the Oslo tunnel all emissions are monitored from a control centre and the air is kept fresh with the aid of 80 powerful fans in each of the two parallel three-lane tubes. These fans start automatically when pollution levels reach a pre-set limit. Dust removal technology with the aid of electrostatic filters has also been introduced. 80 to 90 per cent of particles are removed. A sort of catalytic conversion of exhaust gases from the tunnel is under development. The test results seem rather successful.

Tunnels in loose earth (cut-and-cover) are another possibility used to meet environmental requirements. This is expensive, but in heavily built-up areas, where the traffic is heavy and the surroundings are very noise sensitive, this might be a good solution. It will also reduce the barrier effect that the road produces, if it is covered completely. To make construction cheaper and to reduce other negative consequences, such as pollution and safety problems, cut-and-cover structures have been built in several large cities in the world. In The Netherlands, the experience is that (totally) covering a two x three lane motorway will imply costs of about 50 million dollars per km. In Norway the costs for a total cut-and-cover are estimated to about 10-15 million dollars per km with 2 lanes and 30-50 million dollars per km with 6 lanes. The maintenance cost might be a little cheaper than for a tunnel because the water drainage system can be simpler.

The road can be enclosed in many ways and roads in cuttings can be constructed with all kinds of baffled roof or louvres (see figures IV.5 and VII.3).

The noise at the portal of the tunnel is a problem which has received relatively little attention. The noise level is higher than on a corresponding road in open terrain. The normal level is not reached at less than 50 to 100 metres from the portal. Noise can be reduced by using absorbent materials on the tunnel walls. This has been confirmed in several countries, including Italy, Japan, Austria, Switzerland and Norway. The angle of the portal in relation to the recipient is also important.

Figure IV.4. **Cut-and-cover in suburban area**

Roads on viaducts. The countries that have the most experience in building roads on viaducts to reduce noise are Japan and Italy. Viaducts are an efficient means to reduce noise if taken into consideration at the design stage, and especially if they have solid side panels and no joints, or joints sealed during construction. Screening-off is more effective than in the case of roads built on ground level. The screening effect depends on the barrier height and is related to the distance and height of the observer. In Italy, it has been shown that a noise barrier can be integrated successfully into the design of a viaduct even from an aesthetic point of view. A sort of "safety" barrier can be installed to obtain good acoustic performance as in the case of the "eco-technical" viaduct (see figure IV.6).

Italy has also found that viaducts are more effective in reducing noise if they are designed with a box-shaped cross-section instead of one based on free-standing girders and with narrower expansion joints and solid side walls. Noise-absorbent material can also be affixed to the concrete side walls. If panels are also mounted on the top of the railings or the whole viaduct is clad with transparent material, the noise-reducing effect is very good. There might, however, be reflections from the elements on top, and this must be prevented (see figure IV.6).

The construction costs of viaducts are high and almost the same as the costs for a tunnel. The maintenance costs are, however, much lower because of no expenses for ventilation, lighting, drainage, etc.

From an aesthetic point of view it is difficult to fit such constructions into a built-up or urban area. Several countries claim that viaducts are difficult to build in the light of what the landscape tolerates, and attach varying importance to the integration of the road into the terrain. Some countries (such as Italy and Japan) allow the road to dominate the landscape, while in others (Austria and Norway) the public opinion believes the road should come second.

Figure IV.5. **Different kinds of enclosing road and baffle roofing (reference 1)**

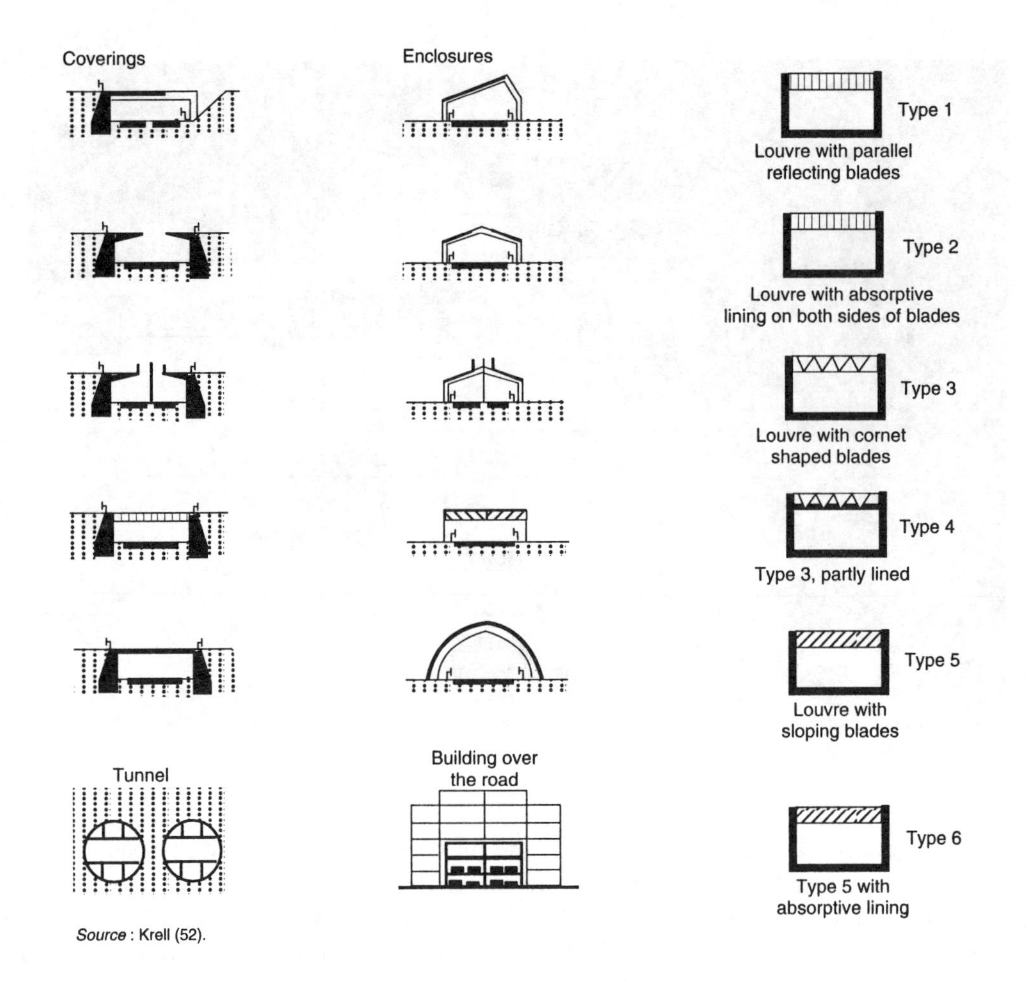

Source : Krell (52).

Use of Roadside Land

The manner in which the road interacts with adjacent land is important for noise control. In some countries, the local authorities have introduced noise control regulations which ensure that no new houses be built in noise-affected areas. In Denmark, for example, road traffic noise is an active criterion in local and municipal planning decisions. No house built after 1980 is subjected to a noise level exceeding the current limits [55 dB(A)]. Some exceptions, however, have been made, especially in densely built-up areas. Standards have been stipulated for acceptable indoor noise and for the lay-out of rooms, so that bedrooms and sitting rooms face the quiet side. The authorities also require that the building acts as a barrier between the road and the outdoor area.

Several countries are aware that more people are being exposed to noise as a result of lack of controls in new development areas and are now endeavouring to achieve better control. In The Netherlands the common requirement is to avoid situating noise-sensitive rooms facing the road. This applies to living rooms, bedrooms and open-plan kitchens.

Buffer zones. Several countries are now attempting to introduce a buffer zone between noise-sensitive areas and motorways or main highways. In Japan these have been established along roads of

Figure IV.6. **"Eco-technical" viaduct constructed in Italy**

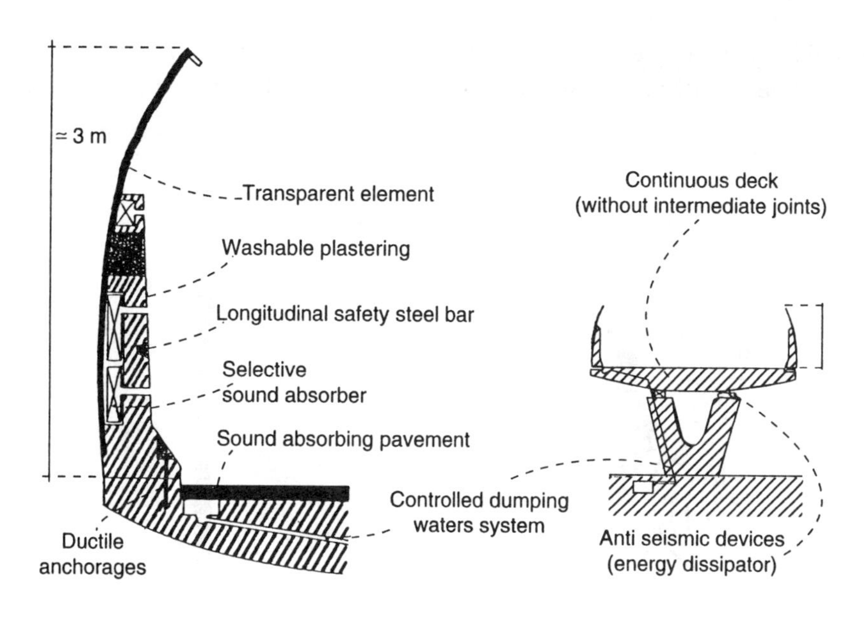

Figure IV.7. **The noise effect with reflective and absorptive shields on viaduct**

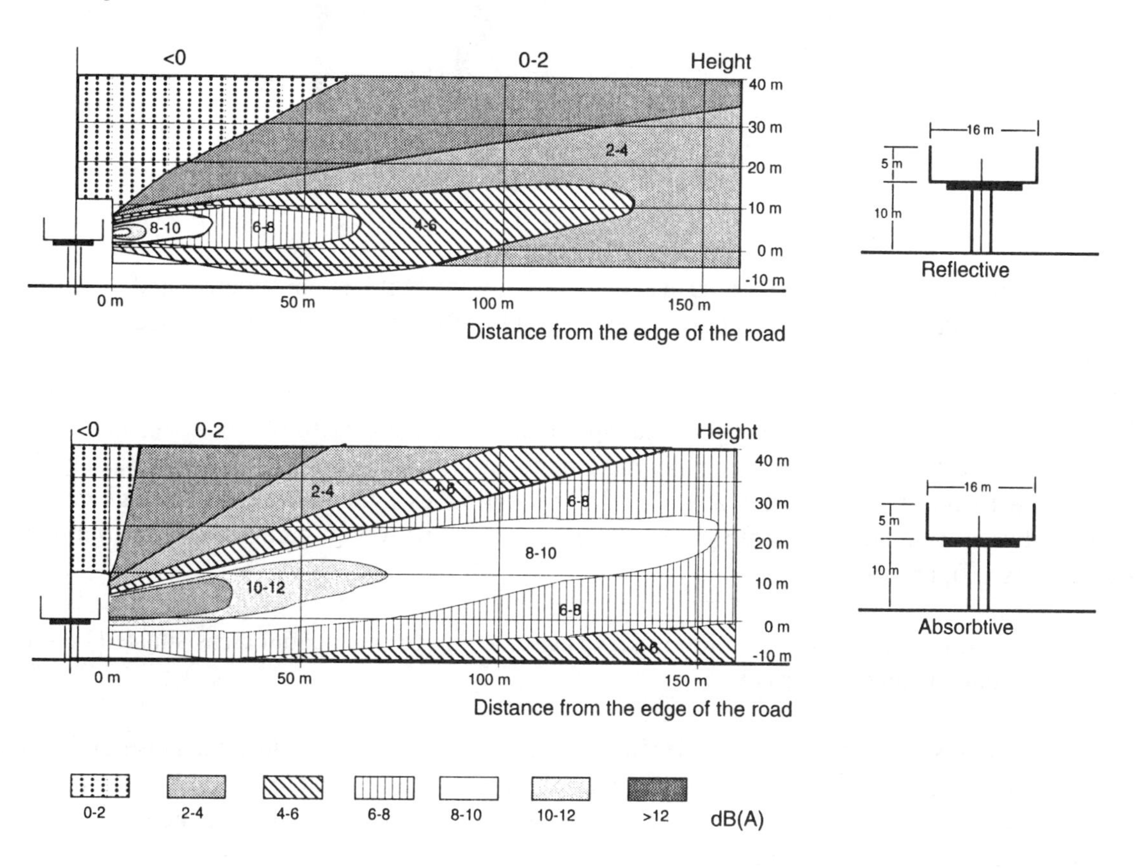

a certain width and along main roads in residential areas. It has often been found to be difficult to reconcile allowing sufficient distance between the road and the built-up area and financial considerations. In a number of countries, including the US, the road authorities have to use state funds to purchase land or development rights in order to prevent new houses from being built along the road. In areas where buildings already exist in the planned buffer zone, the situation is even more difficult. In Finland, houses in new buffer zones are acquired and removed, but these buffer zones seldom exceed 40 metres in width.

Figure IV.8. **Noise reduction by locating building site**

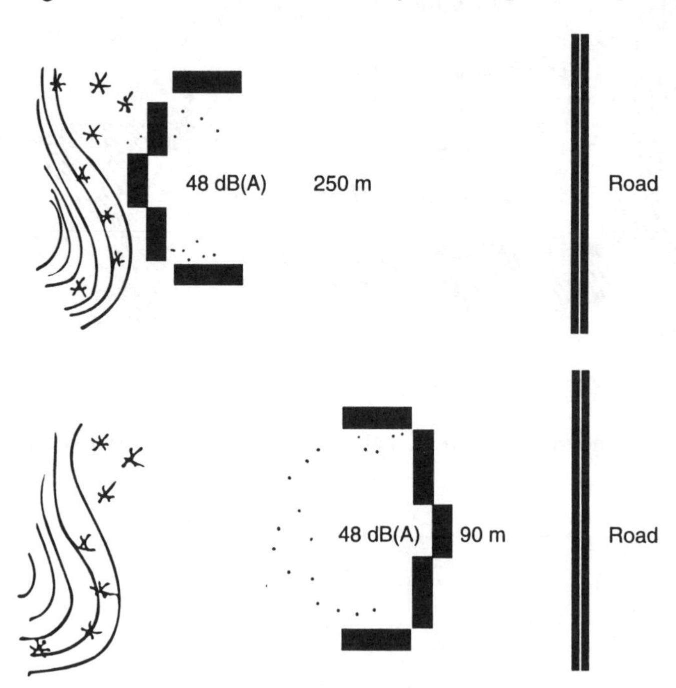

In landuse planning the buffer zone can be reduced, if the building construction and its design take noise into consideration. The building can be a good buffer for an outdoor area (see figure IV.8). Open buffer zones, where only distance moderates the noise level, will often not be large enough in themselves to achieve satisfactory noise abatement. The necessary dimensions of the zone depend on the number of lanes, the traffic and the local speed, and vary from 50 to 600 m. Embankments have proved very effective in buffer zones. A number of countries -- such as The Netherlands, Italy, Japan -- have also found effective buffer zone buildings (continuous / uninterrupted buildings acting as a barrier between the road and the recipient). Often a continuous row of garages or industrial buildings is erected (see figure IV.9).

Office buildings are a likely option to serve as screening for residential areas situated behind, but must be continuous along the road.

Vegetation is another possible barrier in a buffer zone. This might reduce the noise by 3-8 dB(A) but only if the belt is 30-50 m wide and is planted densely both at ground level and higher up. In areas where the wind blows towards the sensitive area, vegetation reduces both wind and noise. Hedges have a very low acoustic effect [0.1 dB(A)] and should not be planted to reduce noise but only to make a visual barrier.

Re-designation of noise-sensitive areas. In zones along the road where the noise level is high and difficult to reduce to an acceptable level by various physical means, it may be necessary to re-designate

Figure IV.9. **Example of a continuous roadside garage as an effective barrier for houses**

the area for another purpose. The change in land use will be realised only after many years if the areas along the road have to be purchased and the buildings acquired.

Improvements Along the Roadside

Most countries do not have legislation requiring improvements in noise conditions along existing roads, although a number of countries have nevertheless given priority to making improvements for buildings. This can be done in several ways and the following describes some of the most common methods:

Noise barriers (see Chapter VI). Free-standing barriers are a common method of reducing noise and should give a satisfactory noise level outdoors. Such barriers have been used in most countries both along new roads and for improvements along existing roads. This is a method that should be used mainly in fairly densely built-up areas, and where the road is located at a certain distance from the buildings, in order to avoid an obstruction effect. When the barrier takes the form of a solid garden fence, for example in densely built-up areas, the barrier will have a positive effect on the areas it is protecting. This type of barrier should be specially designed to harmonise with the buildings.

A noise barrier will protect the area lying in its 'shadow', but in the case of high buildings, the upper floors will not be protected.

Barriers can have different designs depending on where they are erected. The aesthetic and landscape-related requirements may conflict with their acoustic qualities. Where this is the case, some authorities try to avoid using barriers if they give the area an unaesthetic appearance or in order to preserve views.

Green (botanical) barriers may be an alternative. These have proved beneficial in some countries (e.g. The Netherlands), while countries such as Finland and Norway have not found these satisfactory due to the climatic conditions. Vegetation should however always be recommended for its visual effect.

Along roads where noise barriers, etc., are not possible, local screening will be effective in protecting small outdoor areas. This method is relevant particularly in areas where it is difficult to use noise barriers, either for financial or aesthetic reasons. It might also be a good solution for scattered buildings in rural areas. An acceptable noise level might be achieved in areas on the quiet side in the 'shadow' of the buildings.

Facade insulation. Where it is not possible to re-plan room lay-out in existing buildings so that bedrooms and sitting rooms can face the quiet side, facade insulation is an alternative method of reducing noise in homes. This consists of replacing windows, insulating walls and installing ventilation, and is used in densely built-up areas in many countries. Usually, this method will be used when no other measures to reduce noise are possible and, in any case, only satisfies the requirements regarding indoor noise. In the US, schools are the main buildings that obtain noise-reduction by facade insulation, while in most countries in Europe both houses and institutions do so.

By replacing windows with noise-reducing windows (triple and quadruple glazing), the noise level can be reduced to as much as 44 dB(A). The walls and the roof will have to be improved if, as in wooden houses, they have inadequate strength.

Facade insulation is a method that is often used to improve the existing situation. A number of countries (Austria, Japan, The Netherlands, Denmark, Switzerland and Norway) have invested large sums in such improvements. Property owners are often given grants or refunds on their investments. The investment figures and refund percentages vary from one country to another. In most cases it is the road authorities who have invested in these measures; several urban councils have assisted by offering partial financial aid. The reason why these measures are not fully financed is that the residents benefit in other ways from the measures, such as better heat insulation, lower upkeep costs, increase in the value of their property, etc.

The expenses of insulation depend on the extent of the work and how the houses are built. If it is only necessary to change some windows the expenditure will be lower than if the whole wall or several walls have to be insulated, or if ventilation systems have to be installed. How much noise reduction is required is also of importance. The cost might vary from 2.000 dollars per flat up to 50.000 dollars for a one-family house.

The expenses for maintenance are the owners' responsibility. There are no data available.

Most countries recommend a level at 37 dB(A) or up to 40 dB(A) after insulation. Austrian standards give special recommendations concerning noise reduction requirements (see table IV.1) depending on the outdoor noise level.

It is important that houses be adequately ventilated. The drawback here is that the protection from noise is considerably diminished when the windows are opened. Mechanical ventilation is therefore often needed. Ventilators with noise locks have been installed, but experience has shown that these became blocked after a while. Forced or mechanical ventilation has proved to be the best method. In Austria, installation costs are refunded along with the costs of noise-reduced ventilation to ensure fresh air in bedrooms. Ventilation can also be out under the window.

Table IV.1. **Recommendations for noise reducing windows in Austria**

Outdoor noise level	Required noise levels
< 70 dB(A) during day < 60 dB(A) at night	38 dB(A)
> 70 dB(A) during day > 60 dB(A) at night	44 dB(A)

Figure IV.10. **Two different kinds of ventilation systems in the facade**

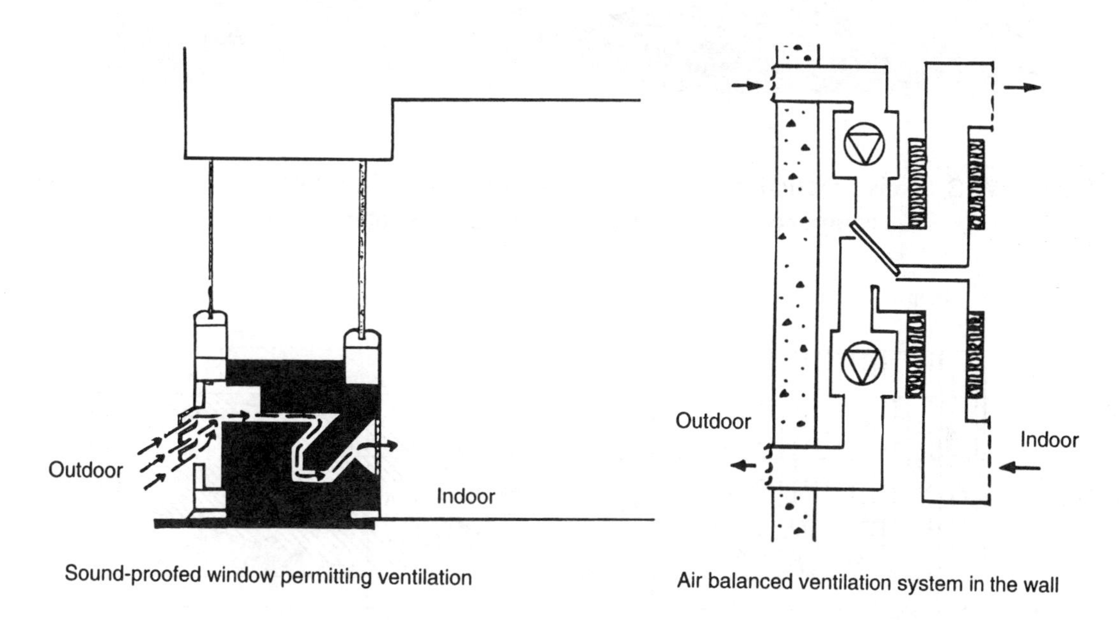

Sound-proofed window permitting ventilation

Air balanced ventilation system in the wall

In some houses, the erection of a new facade (glass veranda or 'gallery') may be an alternative to the complete facade insulation described above. As a rule, this is the best method in the case of total renovation or new buildings. The Netherlands, Denmark and Norway have used this option.

Noise-absorbing materials for different structures. Sound is reflected by concrete walls, supporting structures, etc., along a road. Absorbing material attached to these hard surfaces will reduce noise reflection. The Japanese authorities have mounted sound-absorbent panels with good results on wall surfaces in submerged roads, on tunnel portals and on the facades of buildings. Austria also has some experience in this field. The experience gained from using these methods indicates that good maintenance techniques should be developed and aesthetic standards set.

Figure IV.11. **Facade insulation on an existing house with a glass-veranda**

Figure IV.12. **Noise reduction curves for a retained cut with absorptive vertical walls compared to those without lining (see reference 1)**

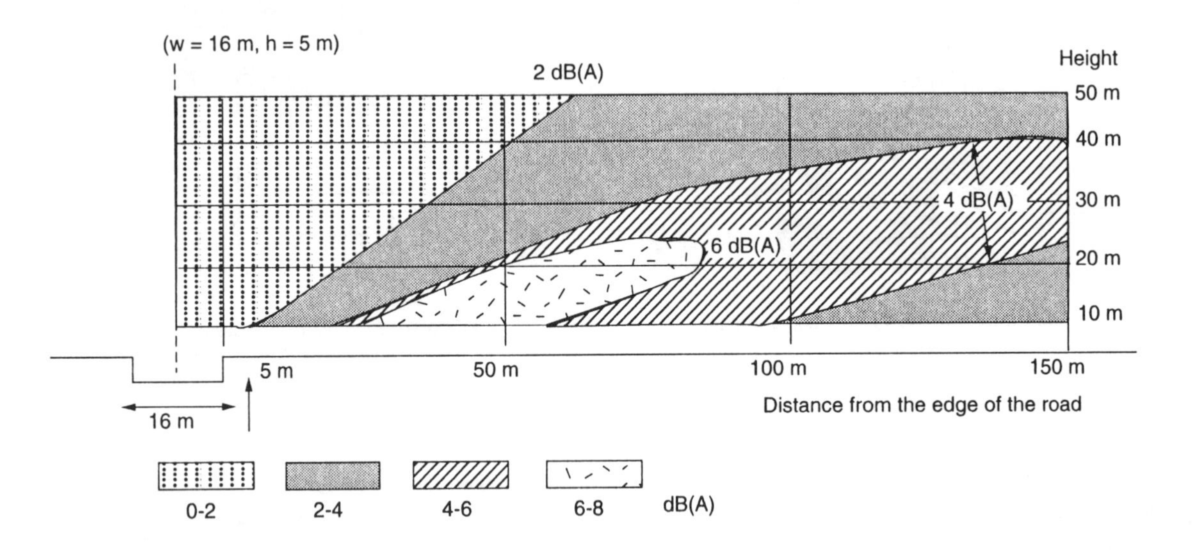

IV.2.2. Traffic control

Traffic control includes, for example, one-way streets, closing roads to vehicles entirely, etc., in order to channel traffic away from or to reduce flow in noise-sensitive areas. Zoning in urban areas can be effective, and ring roads offer the possibility of relieving urban areas, but only if urban planning includes traffic relief.

Heavy vehicles make the most noise. Controlling or directing this type of traffic away from noise-sensitive areas is an effective way of reducing the noise problem. Special routes for heavy vehicles have been introduced in many countries in Europe (such as The Netherlands, Austria), with positive results. In The Netherlands barriers and dams have also been put up along the national trunk roads. The investments are estimated at about 240 million dollars.

Promoting better traffic flow (both smoother and slower) achieved by traffic control, might give a noise reduction of 2-5 dB(A) in urban streets. In The Netherlands, the government has stipulated that local authorities be given contributions for traffic control measures that achieve a reduction of 3 dB(A) or more for houses.

Several European cities have been experimenting for example with financial traffic restraints (tolls on all approaches, taxes, etc.) or by investing more in public transport and in improving conditions for pedestrians and cyclists. This will help to reduce the noise level somewhat, but the effect will only be noticeable when combined with other noise-reducing measures.

Traffic measures and traffic environmental maps are exclusively taken in urban areas they aim at concentrating traffic in zones which are not noise sensitive. This will give the opportunity to reduce the traffic in noise sensitive areas. Such measures are known to reduce the number of people seriously affected. These solutions might also reduce some need for other measures like facade insulation. Improvement costs can thus be reduced perhaps by 30 per cent.

IV.2.3. Reduction of noise at the source

The noise radiating from vehicles is the problem source. At low speeds, engine noise is dominant, while at higher speeds the 'rolling' noise is predominant. Controlling the noise at the source improves the situation for everyone who is subjected to it. The noise can never be eliminated. Other problems caused by road traffic are not solved when noise control measures are directed at vehicles.

Any attempts to achieve a significant reduction in engine noise will require new technology. Even now it is possible to reduce levels somewhat [about 5 dB(A)], especially in the case of heavy vehicles. A noticeable reduction will not be achieved until there is new engine technology introduced. This must be the long-term goal, and international regulations and agreeements will play a role in deciding how quickly it can be achieved. Currently, some countries set more stringent standards than others. In Austria, for example, only the quieter types of heavy vehicles [noise <80 dB(A)] are allowed to cross the Alps at night and speed limits have been introduced (60 km/h) in order to achieve the desired noise control.

To accelerate the replacement rate and bring in less noisy vehicles, a noise-related vehicle tax can be introduced. This will mean that quiet-running vehicles are cheaper.

However, even if engine noise is reduced, rolling noise will remain high. The crucial factor here is the contact between the road surface and the vehicle tyre. Experiments have been carried out in a number of countries with different vehicle tyres and road surfaces in an attempt to reduce this noise (see Chapter V). Particular importance has been attached to the development of noiseless road surfacing. The benefits of more "silent" wheels seem to be potentially high. More research should be done in this area, and new international standards can be set for vehicle tyres (see Chapter VIII).

Table IV.2a. **Summary table of noise reducing measures - Built up areas/urban areas**

	ANTI-NOISE ACTION	Existing roads/buildings				New roads/buildings			
		1	2	3	4	1	2	3	4
I	Roads in cuttings	***	**	****	**	***	****	****	**
	Roads on embankments	**	**	****	**	**	****	****	***
	Tunnels (rock)	****	*	****	*	****	**	****	*
	Cut and cover	****	**	****	*	****	***	****	*
	Viaducts	***	**	****	*	**	***	****	*
	Buffer zone (flat)	***	*	****	*	**	*	****	**
	Buffer zone with barrier	****	**	****	**	****	***	***	**
	Re-design to non noise sensitive area	****	***	***	**	****	**	***	***
	Barriers	**	**	**	***	***	***	**	***
	Local screen	*	****	**	****	*	****	**	****
	Facade insulation	***	***	***	***	***	***	***	**
	Noise absorbent material	*	**	**	***	*	****	**	**
II	Traffic management	**	***	***	***	***	***	***	***
	Special routes for heavy vehicles	**	***	***	***	***	***	***	***
	Smooth flow	**	**	**	***	**	***	**	***
	Increase public transport	*	***	**	**	As for existing roads			
III	More silent vehicles at type approval test	**	**	**	***	As for existing roads			
	Low noise pavement	**	**	*	***	**	****	*	****
	More silent vehicle tyres	**	**	*	****	As for existing roads			
	Control of vehicle noise during use (conformity)	**	**	**	****	As for existing roads			

1 = Efficiency * = poor ; ** = fair ; *** = good ; **** = very good
2 = Feasibility * = poor ; ** = fair ; *** = good ; **** = very good
3 = Durability * = poor ; ** = fair ; *** = good ; **** = very good
4 = Cost * = very high ; ** = high ; *** = medium ; **** = very low

Table IV.2b. **Summary table of noise reducing measures - Suburban/rural areas**

	ANTI-NOISE ACTION	Existing roads				New roads			
		1	2	3	4	1	2	3	4
I	Roads in cuttings	**	**	****	**	***	****	****	***
	Roads on embankments	**	**	****	**	**	****	****	****
	Tunnels (rock)	****	*	****	*	****	***	****	**
	Cut and cover	****	**	****	*	****	****	****	**
	Viaducts	**	**	***	*	**	***	****	*
	Buffer zone (flat)	**	*	***	**	**	****	***	***
	Buffer zone with barrier	****	*	****	**	****	*	****	***
	Re-design to non noise sensitive area	****	**	***	**	****	**	***	**
	Barriers	**	**	**	**	***	***	**	***
	Local screen	*	****	**	****	*	****	**	****
	Facade insulation	***	***	***	****	***	**	***	***
	Noise absorbent material	*	**	**	***	*	***	**	***
II	Traffic management	**	**	***	***	***	****	***	***
	Special routes for heavy vehicles	**	*	***	***	**	***	***	****
	Smooth flow	**	**	**	**	**	***	**	***
	Increase public transport	*	*	**	**	As for existing roads			
III	More silent vehicles at type approval test	*	*	**	**	As for existing roads			
	Low noise pavement	**	**	*	**	**	****	*	****
	More silent vehicle tyres	*	*	**	**	As for existing roads			
	Control of vehicle noise during use (conformity)	*	**	**	****	As for existing roads			

1 = Efficiency * = poor ; ** = fair ; *** = good ; **** = very good
2 = Feasibility * = poor ; ** = fair ; *** = good ; **** = very good
3 = Durability * = poor ; ** = fair ; *** = good ; **** = very good
4 = Cost * = very high ; ** = high ; *** = medium ; **** = very low

IV.3. RECOMMENDATIONS

The choice and use of different noise-reducing measures will naturally vary from place to place, from problem to problem and from country to country. The methods used will depend on the physical situation, financial feasibility, political acceptance and, not least, the cultural values which form the basis for decision-making.

Table IV.2 presents a summary of the efficiency, feasibility, durability and cost of the different kinds of anti-noise measures described in this chapter. The measures with the most asterisks (*) can be regarded as the most suitable solutions.

The common denominator in every case is the need to regard planning and implementation of different measures as part of an overall plan in order to avoid the occurrence of new, undesirable problems. Noise amelioration should take its place in this overall plan to tackle environmental problems generally that the road or the traffic have created or may create.

Roads and their surroundings have different forms and functions. These should be designed in harmony with each other. It is thus also important to take this aspect into consideration when implementing noise-reducing measures. A motorway, for example, clearly has a transport function and is designed for high speed traffic and needs open space around it. Reducing speed or erecting high noise barriers on both sides of the road is not the right answer for this type of road. A street in a town, on the other hand, has a transport function and a local access function, and here the buildings are close to the road. Speed is low. Here the best noise reducing methods will be directed at traffic or streetside buildings. Noise problems do not only occur along the motorway, but also along the approach roads or smaller main roads and especially where the buildings are situated close to the road or street. In these cases, different methods will be chosen to reduce noise. Areas adjacent to roads have different functions or uses. Some are more sensitive to noise than others. An industrial area will tolerate more noise than a residential area or a school. This must also be taken into consideration. The function and form of the road and of its environs must be analysed before measures are implemented to reduce noise.

IV.4. REFERENCES

1. NELSON, P. (1991). *Transportation Noise Reference Book.* Transport Research Laboratory, Department of Transport. Crowthorne.

2. BEYER, E. (1982). *Konstruktiver Lärmschutz, Forschung und Praxis für Verkehrsbauten.* Beton-Verlag. Bonn.

3. VEGDIREKTORATET, MILJØSTYRELSEN (1983). *Prosjektering af boligbebyggelse I støjbelastede byområder, Eksempelsamling.* Copenhagen.

4. EUROPEAN CONFERENCE OF MINISTERS OF TRANSPORT (1990). *Transport Policy and the Environment.* ECMT Ministerial session, Prepared in co-operation with OECD. Paris.

5. NORDISK VEGTEKNISK FORBUND, UTVALG 64, MILJOE (1988). *Vakre veger uten stoyproblem.* Oslo.

6. MARSTEIN, A. (1992). *Basic values for establishing the limits of human toleration of road traffic which vary in different urban areas.* Proceeding of Eurosymposium on "The Mitigation of Road Noise", Nantes, May 12-15 1992. LCPC. Paris.

7. AMUNDSEN, I. and A. MARSTEIN (1988). *Vegtrafikkstoy med hovedvekt på planleggingskriterier for stoyavskjerming på lang sikt.* Public Roads Administration. Oslo.

8. DEPARTMENT OF TRANSPORTATION ENGINEERING AND SINTEF TRANSPORT ENGINEERING (1994). *Efficient and environmentally, friendly freight transport.* Conference proceeding. Vision Eureka, Lillehammer 14-16 june 1994.

9. HANSSON, H.E. (1990). *Design of a composite wheel.* INTROC 1990, Gothenburg.

10. SANDBERG, U. and J.A. EJSMONT (1990). *Tyre/road noise from an experimental composite wheel.* INTROC 90, Gothenburg.

11. CETUR (1981). *Bruit et formes urbaines - Propagation du bruit routier dans les tissus urbains.* Bagneux.

CHAPTER V

LOW NOISE ROAD SURFACINGS

V.1. NOISE CONTROL BY PAVEMENTS

V.1.1. Introduction

The noise experienced by those dwelling near road infrastructures depends, as mentioned earlier, on the acoustic power emitted by vehicles and by any phenomena affecting the sound waves (amplification and/or attenuation) during their propagation toward the receiver.

The acoustic power emitted by the traffic flow depends on the numbers and types of vehicles and the driving conditions. In fact, whether they be heavy or light, vehicles contribute to noise production in different quantities at different frequencies depending on their aerodynamic profile (wind noise), their power-train noise (fan, gearbox and engine, including air intake, cylinder block, exhaust) and their tyre/road noise.

This last type of noise, thanks to the work done by the automotive industry to contain the first two[1], is currently the principal noise emitted by all classes of vehicles, even at moderate speeds, and is the type most closely linked to the nature of the pavement influencing driving conditions. A rough estimate, for urban and motorway conditions, can be derived from figure V.5.

The tyre/road noise is some 2-4 dB(A) greater than the other noises produced by light vehicle traffic cruising at speeds over 50 km/h and by heavy vehicles at speeds starting from 80 km/h.

The tyre/road noise perceived is highly influenced by the road pavement according to the mechanisms shown in figure V.1: there is the action of generation at the tyre/pavement contact point and amplification and propagation actions in which the pavement may have a very important influence.

Separate analysis of these various actions, which result in the noise perceived (i.e. that which disturbs bystanders), has been undertaken to varying extent by the different OECD countries. In this chapter we shall attempt to examine and compare the various results obtained so as to sum up the effects based on the knowledge acquired; they appear quite promising from the standpoint of controlling

---

[1] The noise level of emission has been reduced during the last years by about 10 dB(A) for trucks and 7 dB(A) for passenger vehicles.

roadside noise. The bibliography and references are listed in general and chronological order at the end of the chapter.

Figure V.1. **Composition of received global tyre/road noise**

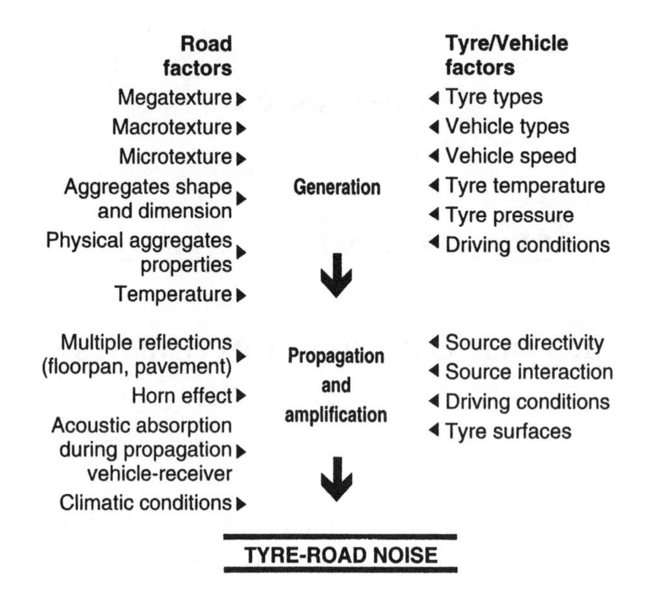

V.1.2. Generation of tyre-road noise

The mechanisms generating tyre-road noise are strictly connected with the surface characteristics of the pavement, which interacts with the tyre as summarised in figure V.2 under the terms: radial vibrations, air resonances, adhesion mechanisms.

Figure V.2. **Tyre-road noise generation mechanisms**

Radial Vibration	**Air Resonance**	**Adhesion mechanism**
♦ Impact of tyre tread blocks on road surfaces	♦ Pipe resonance ♦ Helmholtz resonance ♦ Pocket air-pumping	♦ Stick-slip motions --> tangential tyre vibrations
♦ Impact of road surface texture on the tyre tread		♦ Rubber-to-road stick-release

Radial vibrations are caused by irregularities of the rolling surfaces having wavelengths close to the mean radius of the contact zone, which is influenced by the mega-texture of the pavement (50-100 mm)(see figure V.4).

These vibrations, transmitted via the suspensions, excite resonance phenomena both inside the cabin -- which reflect in this way the irregularities in the megatexture of the pavement -- and outside, between the pavement and the underbody of the vehicle.

Air resonances are produced by more complex mechanisms involved in tyre-pavement contact, and occur in the medium and high frequencies.

It is assumed that the third generation mechanism is linked to tyre/pavement "adhesion", and includes the contact/release sequences (cycles) of the two surfaces, as well as the action of actual adhesion and detachment of the single elements of rubber and stone. These mechanisms generate tangential vibrations of the tyre. The noise generated by these mechanisms is shown in figure V.3 as source S1.

Figure V.3. **Tyre-road noise propagation mechanisms**

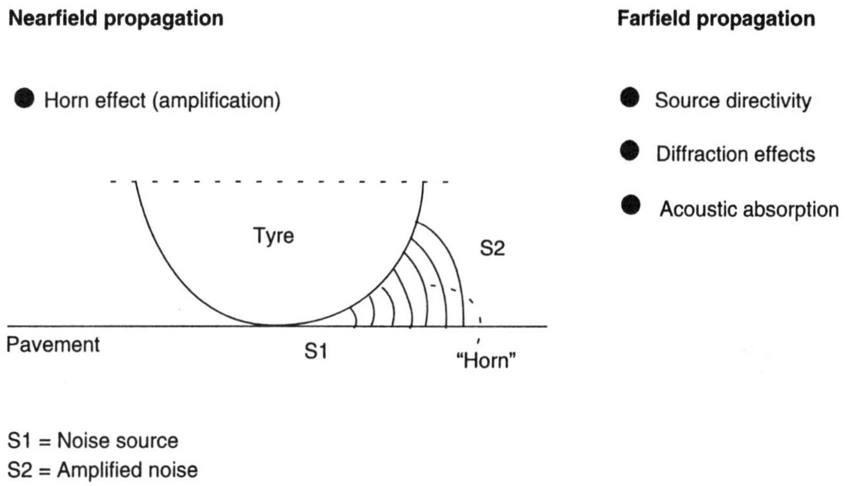

Nearfield propagation

- Horn effect (amplification)

Farfield propagation

- Source directivity
- Diffraction effects
- Acoustic absorption

Tyre

S2

Pavement

S1

"Horn"

S1 = Noise source
S2 = Amplified noise

The spectral components of the noise emitted at S1 are located in both the low and high frequencies of the spectrum. Therefore, as the energy in the high-frequency range predominates, it is a decreasing function of the irregularities in the pavement, as shown in figure V.4.b.

V.1.3. Tyre/road noise propagation

The effective noise propagated to the outside, however, is that indicated as S2, i.e. that which is generated as S1 and then amplified by the "horn" effect (also called the "trumpet" or "dihedral" effect) and the other mechanisms listed in figure V.3. This will then be perceived, attenuated, by the receivers surrounding the road.

The "horn" effect is the reflection of the waves from the walls of the dihedral formed by the surface of the tyre and the road surfaces both ahead of and behind the contact area. During propagation from the sources to the receiver, the noise can be controlled by means of sound absorption properties of the pavements. Three distinct effects can be identified:

- Reduction of horn-effect amplification, and absorption of noise radiated from mechanical sources;
- Reduced reflection between the vehicle underbody and the road surface;

Figure V.4. **Influence of superficial and absorbing caracteristics on noise**

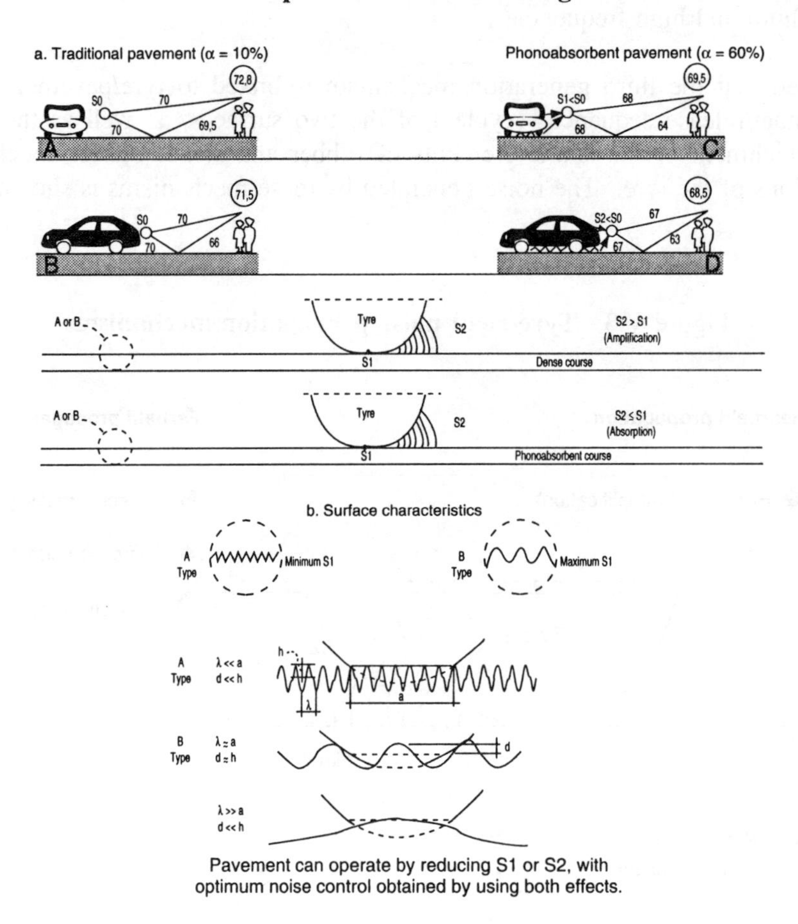

- Absorption of sound waves generated by tyre-road noise and propagated over the terrain between the vehicle and the receiver.

At this point, one can better understand how pavements can influence tyre-road noise and also the other noise generated by traffic. They can affect the generation of tyre-road noise (S1) by vibrations induced in the tyres by the roughness of the road (in the range of mega- and macrotextures) and its amplification (S2) by air resonances (the "horn" effect); they can also affect all noise generated, through global or local absorption (figure V.4): global absorption refers to the attenuation of all sound waves generated by all sources, reflected several times between the vehicle underbody and the road surface (multiple reflection), while local absorption is due to attenuation of the "horn" effect and acts on S2. Both types of absorption are due to the surface and internal structure of the pavement. The sound waves striking these absorbing structures are attenuated by multiple reflections within the residual voids they contain.

V.1.4. Low noise road pavements

Since the end of the 1970s, as a result of the development of methods for measuring sound energy and comparing the performance of pavements in this regard, knowledge of the phenomena producing these effects has steadily grown, leading to the recent adoption, in practically all OECD countries, of various measures, related to pavement design, materials, and application techniques, aimed at reducing tyre-road noise.

In addition to better tyre performance, the diversification of pavement surfacings, implemented by Road Authorities since the middle of the 1980s, has made it possible to obtain a considerable reduction in noise generation and amplification.

Road pavements thus provide, to an ever greater extent, the bearing capacity, skid resistance, and riding comfort necessary for driving safety, but now also function in an optimal manner to reduce noise generation. A road surface can be chosen by comparing performance levels achievable with different micro-, macro-, and megatexture characteristics, obtained using economically diversified mixtures, with functions such as drainability and safety required of the pavement in the design stage.

All countries have come a long way to reach the solutions described below; all of them started from the comparative analysis of the noises generated at the roadside by the various types of existing pavements (concrete, bituminous, concrete block pavements), using methods to compare their acoustic performance that, as mentioned above, will be described briefly in the next section.

Figure V.5 shows the noise levels measured at the roadside and inside vehicles for the various types of pavements assessed. Rather than specific values, the various ranges reported for the different OECD countries are given. The variability arises quite naturally from the lack of homogeneity of the noise sources, as well as the extreme structural diversity of surfaces until now called by the same name. What is important is to evaluate the mutual positions of the ranges, especially those of the noise measured at roadside, which takes into account (in some cases) the effects of global and local absorptions. It can be seen that porous asphalt pavements (also called draining or sound-absorbing) are among the most valid from the acoustic standpoint.

Figure V.5. **Typical range of tyre/road noise levels according to the type of road surface**

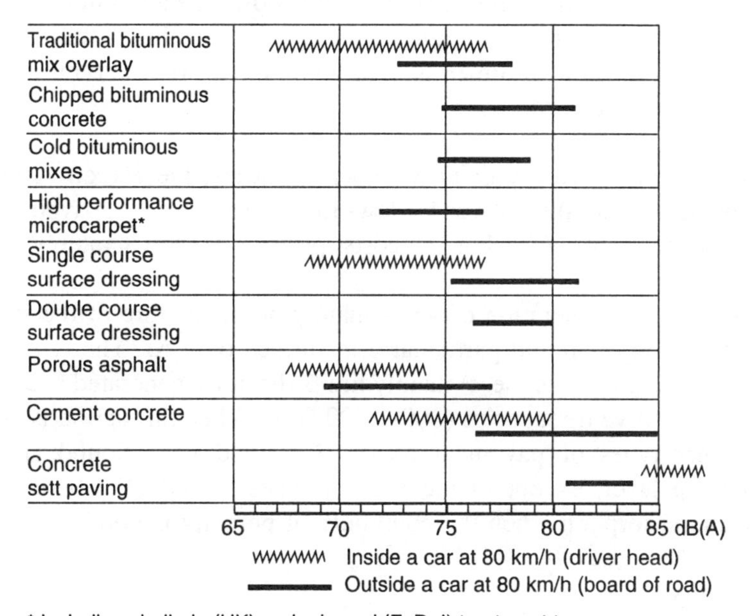

There is also reduced noise on fine and regular surfaces or where the mega- and macrotexture values are low, thus reducing local deformations of the tyres. This effect can be obtained by using small aggregates with open gradings, and micro- and thin carpets. This type of pavement, while it reduces the noise induced by tyre vibrations, tends at high speed to increase the component due to air resonances (it has a negligible sound-absorption effect), but still has high skid resistance values.

Draining asphalt was the first road pavement technique identified as a low-noise surfacing, and acoustical research has been devoted mainly to optimising its effectiveness. From the very beginning, it has been used to reduce noise generation and attenuate any noise produced. The global effectiveness of draining pavements is strongly influenced by their sound absorption properties as well as by the macrotexture characteristics resulting from aggregate sizes and the laying process. Tyre-road noise due to vibration at low frequencies can be further attenuated by reducing the macro-roughness during the rolling stage.

When porous asphalt surfaces are laid, the degree of success in obtaining "smooth" aggregate surfaces (through flattening during the rolling stage) and low macro-roughness will determine the extent of reduction of rolling noise caused by vibration of the tyres at low frequencies and of noise inside the vehicle. In addition, the degree of porosity obtained by mix design and laying will determine the reduction of the air resonances that produce air pumping, with the above-mentioned effects of a reduction of noise emission at high frequencies and global and local absorption of any noise emitted, as well as a highly desirable elimination of tyre spray in wet weather.

Many studies have shown that the long-term conservation of porosity depends on the performance of the materials (aggregate grading, type and percentage of binder, etc.), on traffic conditions (speed, types and percentages of vehicles, etc.), and on maintenance operations (unclogging process, upgrade of lateral drainage). Despite their acoustical effectiveness, draining asphalts can not be used universally to mitigate traffic noise, because some aspects of level of service (durability, clogging, winter maintenance) will sometimes pose limitations; in such cases it is possible to use other low-noise surfaces, such as "thin carpets".

Recent studies and applied research are also adding to our knowledge of the behaviour of micro-carpets with respect to traffic noise. These should allow a better compromise between emissions at low frequency and at medium and high frequencies, achieving a micro-texture with surface irregularities on the order of a few millimetres, which do not deform the tyres and reduce the pressure of the trapped air but still ensure good skid resistance[1].

In short, based on input from OECD Member countries, the Expert Group has prepared the following chart showing various alternatives for low-noise pavements, i.e. pavements that reduce and/or absorb tyre-road noise (see table V.1). For the costs, please refer to Chapter VII.

As has been seen, noise generation depends mainly on the texture of the surface. The ability to damp this type of noise (S1) is found particularly in surface type A. The type A pavement with the "microtextured surface" (2-3 cm thickness) mainly limits the noise generated and, when pervious, is not absorbent at the most sensitive frequencies (800-1 200 Hz). When this action is prevalent, on the other hand, there exist other types of pavement (types B, C and D). Sound absorption is generally concentrated in the top layer, except in the case of types C and D, where the lower layers also contribute some sound absorption when the continuity of porosity is good.

A-type pavements

The type A pavement, used in Germany (Stone Mastic Asphalt-SMA), France (0/6 Béton Bitumineux Très Mince - BBTM and Béton Bitumineux Ultra Mince - BBUM), and Italy (thin carpets and grip-road), is produced by simply spreading an overlay having a suitable surface microtexture over

---

1 Several countries (I, D, F) provide for the use of micro-carpets as overlays on porous pavements that have become slippery, thus still partially maintaining sound-absorption.

Table V.1. **Surfacings with low tyre/road noise**

Type of structure ans anti-noise effectes	Anti-noise elements	Additional functions	Country	Sound absorption coefficient (α) in different frequency bands (Hz) <700 ; 700-1250 ; >1250	Construction problems	Management observations	Notes
A. Microtextured surfacing Traditional (black/white)) 2/3 cm (max) Low S1, excitation of vibration	• Fine and ultrafine bituminous wearing course (e.g. micro-carpets) • Surface treatment (e.g. griproad, micro-griproad and spraygrip) • Exposed aggregate concrete[1]	High skid resistance	D, F, UK, I, S, A	0 0.1 0.1 0 0.2 0.35	To be laid on smooth pre-existing pavement	• Complete information lacking on durability and more information needed on acoustic performance • Limited costs if microcarpets are layed at regular frequency	• Good for use on urban streets and especially on ring roads and by-passes • Sound absorption characteristics can be obtained using porous aggregates
B. Macroporous surfacing ; medium thickness Traditional 3/8 cm Low S2 (sound absorbent)	• Drainage asphalt as wearing course (porosity ranging from 20 to 28%) mixed with normal bitumen or preferably polymer modified bitumen	High permeability which can be maintained also on high speed roads	All European countries	Good absorption at medium frequencies (500-1200 Hz) depending on thickness 0.2 0.45 0.60	To be constructed with impervious and regular subcourses	• Acoustical efficiency over time if porosity does not deteriorate rapidly • Average costs • Good drainability with residual voids > 20% • To maintain permeability, cleaning operations are sometimes required (preventive or curative unclogging) • During winter operations special treatment and de-icing salts are required	• Widespread use on high speed roads • Acoustic effects can be enhanced with use of barriers • Used in road sections with cross profile inversions
C. Macroporous medium and high thickness up to 50 cm Traditional Low S2 (sound absorbent)	• Bituminous or cement concrete in one or several courses (porosity increases with depth)	Good performance in tyre noise reduction, but generally used when water disposal effect is sought	D, F	0.2 0.5 0.65 (15 cm) 0.2 0.5 0.65 (50 cm)	Need of drainage system at bottom (near the subbase)	• Valid over time • High initial costs • Need complex system of deep drainage	• Used when strongly justified by need to reduce size of drainage system (due to high costs)
D. Euphonic 4/6 cm Concrete slab Low S2 (sound absorbent)	• Porous asphalt and continuously reinforced concrete with resonators	• High durability • High evenness • Conceptually efficient over entire frequency range • Can substitute barriers and covers in urban areas	I	0.60 0.45 0.45	Need of drainage system at bottom (near the subbase)	• High initial costs	• Experimental stage • Conceptually very good for heavy freight traffic

[1] Low noise cement concrete pavements can be obtained using small size aggregates in the surface layer and by simple construction techniques, and also with specific treatments such as griding or covering with epoxy resin surface dressings, or a new course of exposed aggregate concrete.

Remark: For B, C, D types, it is possible to intervene on S1 noise level a traditional flexible, semirigid, or rigid pavement. This is especially well suited to urban areas, or roads with limited freight traffic that generate noise at the lower frequencies. This type is characterised by either nil or reduced porosity (residual voids < 2 mm) and the mixtures are open graded. The type A pavement can also be made up of a surface layer of the "grip-road" or "shell-grip" type, containing a percentage of porous aggregates with small but very strong projections, or else with thin or ultrathin carpets.

For this pavement, the minimum values of the sound absorption coefficient (the measurement methods for which are discussed in the next section) are 0.0, 0.1, and 0.1, respectively, for the three frequency ranges of 0-700 Hz, 700-1 250 Hz, and 1 250-2 000 Hz, except in the case of mixtures made up of porous and open-graded aggregates, where the values rise to 0.0, 0.2, and 0.35.

B-type pavements

The second type of pavement, type B, the most widespread in all OECD countries (but with widely varying percentages of residual voids), is termed "medium thick macroporous" (3-8 cm thickness) because it is formed of a surface layer of draining/sound absorbing asphalt mix produced with either normal or polymer-modified binders with residual voids on the order of 1-10 mm constituting not less than 20 per cent of the volume of the layer (internal macroporosity).

Porous asphalt mixtures dampen noise in three ways:

1. By minimising the generation of tyre-road noise by:
   - Cutting vibrations by flattening surface aggregates during laying;
   - Reducing air resonance phenomena through porosity, so preventing pressurisation of entrapped air;
2. Attenuation of amplification by the tyre-pavement horn effect; the porous asphalt pavement limits this effect;
3. "Excess" attenuation, or the difference in sound propagation between a reflecting pavement and an absorbent one.

In addition to its acoustic and photometric properties, this type of pavement is also particularly effective in eliminating rain water (reducing the risk of aquaplaning) and spray.

For this type of pavement the sound absorption coefficient has good values (up to 0.8) in the range of 500-1 200 Hz, depending on the thickness. In fact, for the three frequency ranges of 0-700 Hz, 700-1 250 Hz, and 1 250-2 000 Hz, it reaches the minimum values of 0.00, 0.20, and 0.55, which may rise to 0.2, 0.45, and 0.6 depending on the thickness and the quantity and quality of the intercommunicating voids, as well as on the relevant surface aggregate grading, which influences the S1 noise produced.

The B-type pavement is chosen by road managers to reduce the potential hazard of aquaplaning. Under traffic, porous asphalt surfaces become clogged, significantly reducing their draining capability and also their acoustical performance. Other management problems will be mentioned in section V.3, together with the solutions for dealing with them.

C-type pavement

The third type of pavement, type C, studied and tested full-scale in Germany, and more recently in France, is "medium and very thick macroporous" (up to 50 cm thick). For this type of pavement, in which the noise-reducing elements consist of one or more courses of bituminous or cement concrete, the sound absorption coefficient depends on the thickness, since the thickest pavement can be effective over the whole frequency range (100-5 000 Hz); on the other hand, this type of pavement is used mainly for its water disposal capacity.

For porous concrete pavements, the first prerequisite for low tyre-road noise is an optimal surface geometry; this entails the use of a longitudinal smoother and aggregates not larger than 11 mm, or even better 8 mm. Otherwise, more tyre-road noise will be generated than the pavement voids can absorb.

It is fitting to insert here certain observations regarding concrete pavements in general. Besides what can be achieved with the use of porous concretes, considerable reductions in tyre-road noise can be obtained nowadays with relatively simple construction techniques and surface treatments, both aimed at optimising pavement surface geometry. New materials are available for this purpose, mainly consisting of concretes with reduced maximum aggregate size, polymer modified mortars and epoxy coatings.

D-type pavement

The fourth type, D, is the "euphonic" pavement, an improvement of the polyfunctional composite pavement recently developed and introduced on the Italian motorway network, which consists of a surface layer of porous asphalt (with draining and sound-absorbing characteristics) some 40-60mm thick overlying a continuously-reinforced concrete slab with resonators of about 500 cm3 each, distributed over the entire road surface in the manner shown in cross section and plan in figure V.6.

For this pavement, the sound absorption coefficient for the three frequency ranges of 0-700 Hz, 750-1 200 Hz, and 1 250-2 000 Hz reaches values of 0.6, 0.4, and 0.7, obtained by a combination of the effects of the top layer and of the cavities, which naturally influence the absorption of the lower frequencies (see diagram in figure V.6).

Studies and scale models of this "euphonic" type pavement have been completed, and it is shortly scheduled for implementation on the road. Full-scale experiments will certainly show what performance can be achieved in practice given all the construction problems. Nevertheless, it is hoped to obtain a more constant sound absorption at the different frequencies than was achieved in the experimental stage.

Since, as has been noted, most of the noise generated by traffic is related to the rolling of the tyre on the road surface, there is justifiable interest in solutions that can reduce this effect; it has in fact been shown that the overall attenuation obtainable can be equivalent to a reduction of 50 per cent in the mean traffic speed or to a tenfold reduction in traffic volume, or to the installation of acoustical barriers 2.5m high. In particular, in addition to reducing tyre-road noise, the best anti-noise surfaces and/or pavement structures can also reduce exhaust noise by 2 dB and motor noise [up to 5-6 dB(A)], with the most marked attenuations occurring in the medium- and high- frequency ranges, from 500 to 1 600 Hz.

It is clear at this point how important it is to have pavements with better sound absorption coefficients at the lower frequencies, since a reduction of 5 or 6 dB(A) at these frequencies approximately corresponds to the effect one would have by doubling the surface mass of the facades

Figure V.6. **Sound absorption coefficient for euphonic pavement**

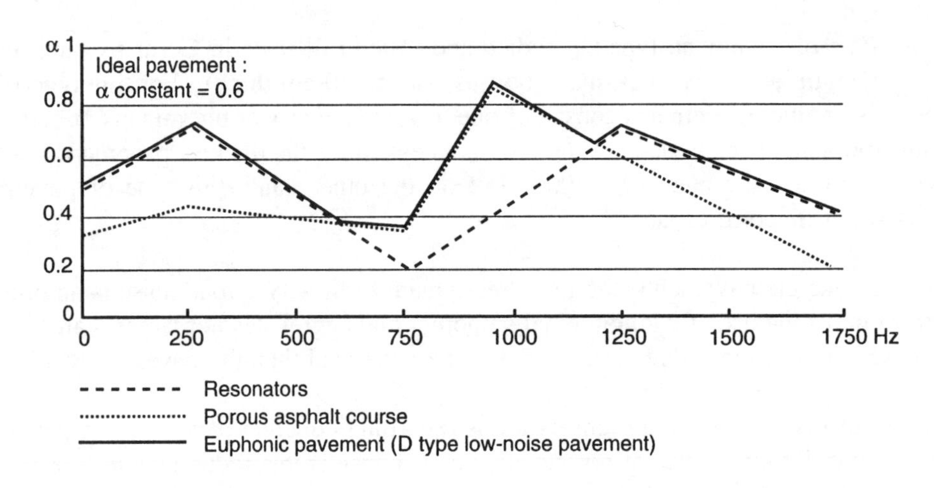

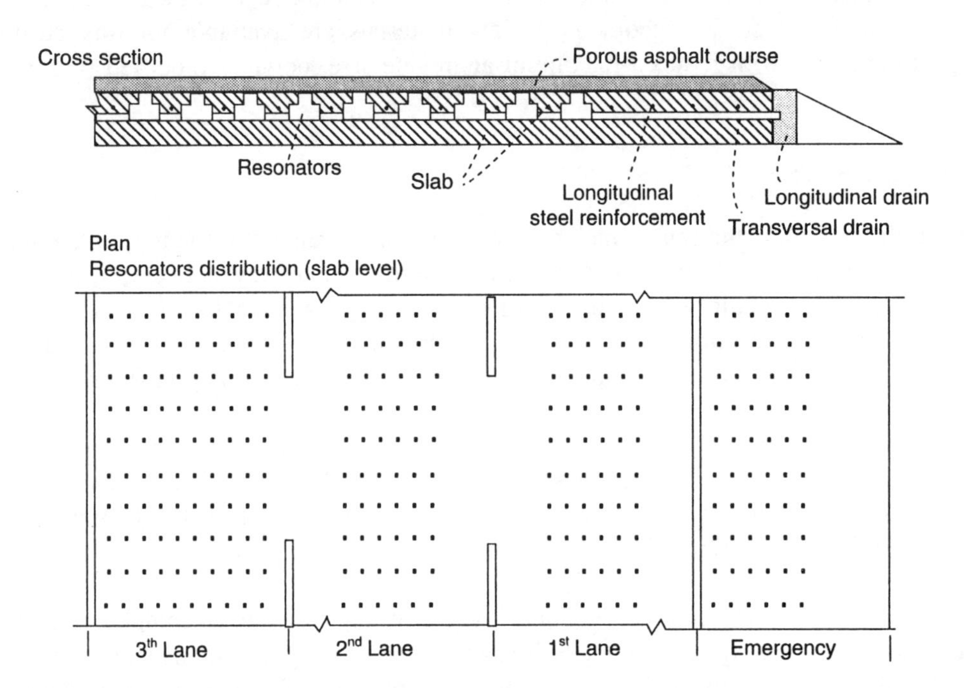

or multiplying the distance between the disturbed building and the road by four. For this reason, it is far more important to step up research and development on type D pavements with increased sound absorption, currently the most promising solution to the problem.

V.2. METHODS OF MEASURING TYRE/ROAD NOISE

Besides providing a summary description of the noise measurement methods used and their limitations, this section will also set out briefly some of the experiments conducted by the OECD countries most interested in controlling traffic noise through the optimisation of pavement surface characteristics.

The method used to measure the global noise emitted by a vehicle has been standardised by the ISO (International Standards Organisation) in Standards 362 (1981) and 7188 (1985), which refer to the conditions of maximum emission in an urban area. The method is as follows: on the road, using a microphone positioned in pre-arranged manner, under controlled weather conditions, two free-range sound measurements are taken from both sides of the noise of a vehicle accelerating rapidly from a specified speed (50 km/h according to ISO 7188). As mentioned above, the automobile industry, using this method, has been able to modify vehicles so as to limit their contribution to the noise pollution caused by traffic, in keeping with the relevant UNO/EEC regulations. This has made it possible to limit the noise produced by the propulsion system at low speeds.

Existing law requires limiting overall noise, including tyre-road noise; this is being implemented by harmonising existing national regulations or recommendations regarding related measurement methods developed subsequent to the above-mentioned ISO methods, through the measurement of noise with the vehicle launched and with trailer. The purpose of the current studies and research is to develop a global survey method that can distinguish the traffic noise component attributable solely to the pavement, in order to develop acoustic quality control of road pavements and so be able to classify them acoustically on the basis of scientific-technical considerations valid in any environmental conditions rather than of empirical criteria.

Figure V.7. **Sound absorbing measurements with RIMA machine**

Of the many methods used to measure tyre-road noise, those used in OECD countries and worth mentioning are: the statistical "pass-by" method (UK, D, F, DK, N, S, I), the "controlled pass-by" method (F and D), and the "trailer" method (A, P, D, S, E, I, FIN). This last method is combined with global sound absorption measurements. One should also mention on-road sound absorption measurement employing a standardised sound source, used in Italy, mainly to classify pavements and check performance (see figure V.7). The current ISO Working Group has retained only the SPB method; it has been shown that the trailer method together with the sound absorption coefficient can be correlated with the SPB method for light vehicles.

It is precisely the contribution of these latter methods, developed recently, concurrently with the growing use of porous pavements, that has established a basis for a rational approach to control of traffic noise by means of pavement surfaces (and not only porous ones), in which it has been demonstrated that no single measurement procedure can respond to all the various needs.

The noise due to the passage of vehicles on a pavement is thus measured in dB(A) on the roadside, so determining the noise emitted by the traffic, as is or under specified transit conditions, or by means of a measuring trailer, with or without pavement absorption measurements; these last two procedures determine the tyre/pavement interaction noise independently of the noise of the engine, exhaust, or other vehicle parts (see table V.2 on methods for measuring noise emission on pavements, with the advantages and disadvantages of each measurement technique).

Table V.2. **Methods for measuring tyre/road noise**

METHOD	Measurement of total noise of vehicle passage (statistical pass-by)	Measurement of total noise of selected vehicle passage and tyres (controlled pass-by)	Measurement of tyre-road noise (trailer)	Measurement of sound absorption
MEASUREMENT CONDITION AND LOCATION	Laterally to road and in open field condition	Laterally to road and in open field condition	On road surface at tyre contact point	In laboratory and on road surface
TYPE OF NOISE	Representative of actual traffic (passenger and freight vehicles)	Representative of two light vehicles and four type of tyres	Representative of a reference tyre or of 4 or 5 different types, depending on type trailer	Impulsive type noise emitted close the pavement
RESULTS OF MEASUREMENT	Noise level at a definite speed determined by regression	Noise level at a definite speed determined by regression	Noise level at a definite speed determined by regression	Absorption coefficient at different frequencies
ADVANTAGES	Measurements permit determination of noise perceived off road	Measurements permit classification of road surfaces with respect to light vehicle traffic	Measurements permit determination only if tyre-road noise, but without preventing consideration of local absorptions	Method is necessary as complement to tyre road noise measurements (trailer)
DISADVANTAGES	• With limited number of samplings, influence of type of traffic is considerable • With high number of samplings effect of traffic type is reduced but onerousness of samplings increases • Can not be used to classify road surfaces (a large number of factors influences this measurement)	• Laborious nature of method due to need to have two lanes free of traffic • Does not consider the tyre road noise caused by freight traffic	• Does not permit consideration of global absorption • Normally used to measure only tyres of passenger vehicles and thus does not take into account tyre-road noise from freight vehicles	• Complementary type measurements to define pavement noisiness
COUNTRY	UK, D, F, I, DK, N, S	F, D, I	A, P, D, S, E, I, NL, FIN	I, F

In the first case, two distinct measuring procedures are envisaged. The first, known as "statistical pass-by", is so termed because the measurements are made on a statistically significant number of vehicles representative of the vehicle population actually in circulation, possibly subdivided into predetermined categories (such as light, medium and heavy), as well as the cruising conditions or respective speeds.

This procedure entails installing microphones for long durations (sufficient to determine constant flow conditions) and measuring noise levels (for example, sound pressure) and/or the noise frequency spectrum, and simultaneously recording the speeds and types of vehicles passing (for example, the percentage of freight vehicles).

The noise emitted by a given category of vehicle is customarily represented by the S.E.L. or the Lmax.

This method, used in several countries because of its ease of execution and the limited number of technicians required, nevertheless has certain drawbacks, set out below. In addition to the difficulty of generalising to road networks that are heterogeneous in terms of traffic and pavement types, the method requires repeating the measurements over time, with many determinations each time, to allow for the natural evolution that can occur in the vehicle population and in tyre characteristics. Also, one must bear in mind not only the effects of the possible presence of safety barriers or unpaved and absorbing surfaces on the measurement itself, but also more especially the difficulties of correcting for the vehicle type and tyre type when pavements are classified using this method.

The second procedure for measuring traffic noise is called the "controlled pass-by" and entails the use of selected test vehicles, fitted with typical tyres, passing at known speed. This method was developed as a variant of the "coast-by" and "drive-by" methods during the cooperative test campaign conducted in France and Germany, and has since been widely adopted. This method, which requires the selection of a number of particularly representative vehicles and tyres, has several practical drawbacks when used on highway sections with heavy traffic.

The third type of traffic noise measurement, the "trailer" method, is a mobile survey; it quantifies only rolling noise. The measurement is made using a trailer having its wheels enclosed in an anechoic chamber so as to eliminate the influence of background traffic noise. Within the chamber are several microphones near one or more wheels, providing one or more reference tyres. Continuing a project of the University of Stuttgart, various European laboratories have conducted specific experiments using this measuring system.

In Austria, in particular, the measurements, never made at temperatures below 10°C, are performed at a speed of 100 km/h using a trailer having a microphone, placed at a height of 10 cm, 40 cm behind the centre of a measuring wheel. The measuring wheel is fitted with a special type of tyre (PIARC 165 R, at 2.3 bars pressure, loaded with 400 kg).

In Finland, recent rolling noise measurements have been made at speeds ranging from 60 km/h to 120 km/h (in 20-km/h steps), using a system installed on a trailer having a reception device active for 10s, with studded tyres, the type used most frequently during the long winter in the Nordic countries.

In Spain, the measurements are made continuously using a single-wheel trailer, specially insulated from outside noise and equipped with three directional microphones 50 cm from the vertical of the single wheel used for the measurement; the use of a single wheel eliminates any influence of noise from other wheels on the trailer. The frequency spectrum obtained from these recordings is analyzed every 100 m, and in this way it is possible not only to compare different pavements with typical reference tyres, but also to test the latter on a reference road surface. A few researchers have compared the values obtained with the trailer and those obtained from previous measurements so as to correlate the two types of measurement.

The results yielded by the trailer method might lead us to underestimate the sound absorption provided by porous pavements; moreover, the pavement classification based on these determinations may be influenced by the type of tyre selected in the case of the single measuring wheel.

Because of the fact that, for certain road surfaces, it is essential to know the sound absorption value in order to determine the level of noise emission, it should be noted that it is also possible, with the trailer method, to take account of pavement absorption in the measurements obtained, using a constant that depends on the type of tyre used.

Finally, it is worth noting that the values obtained by application of the principal measuring methods described above correlate very well with one another if the same type of tyre is used. In fact, sound emission may differ on the same surface if the tyres used for measurement are different; for this reason the trailer method is being developed to use 4 or 5 measuring wheels.

To allow for the sound-absorbing properties of the road surfaces, or the acoustic absorption coefficient, which is the fraction of energy absorbed when a sound wave is reflected, both traditional laboratory measurements and on-road measurements have been used.

These measurements made it possible to determine the values of the aforesaid coefficient as a function of the frequencies of the sound spectrum considered at the angle of incidence of the actual wave. In this way, it was found that the absorption coefficient curve is characterised by a series of resonances at frequencies that are in an arithmetic progression, and that the frequency and amplitude of the maxima and minima are related to physical characteristics of the pavements, such as thickness, porosity, flow resistance, and the shape factor.

On this basis, it was also possible to develop a scientific and not merely empirical approach to resolving the problem of optimising the acoustic properties of the porous mixes used in constructing sound-absorbing pavements.

The laboratory measurements were made at normal incidence using both the standing-wave tube (the "impedance" or "Kundt" tube), with calculation of the sound absorption coefficients between 100 and 800 Hz, and the reverberating chamber, which measures the sound absorption in diffuse sound field conditions.

As far as on-road measurements are concerned, the impulsive technique developed in France has the same purpose as the laboratory methods. They permit to determine the excess attenuation, under normal incidence or near the ground function of the distance. In this case, the excess attenuation represents the ground effect which can be added to the spherical spreading.

In addition to these procedures and techniques, another recent development, in Italy, is "RI.MA." equipment, an automatic, vehicle-mounted measuring device that can perform static surveys on the road, using the above-mentioned impulsive technique and taking account of the global effect of the pavement on road traffic noise emission. This technique consists of generating a signal of brief duration from a special source (excitation loudspeaker) and using microphonic probes to record the signals related to the direct wave, the reflected wave, and the waves diffracted by other obstacles present. The method thus provides a series of spot measurements even in the presence of traffic, simply and rapidly (about 90 seconds per measuring point), referred to a fixed emission spectrum (sound source), thus allowing comparison of results obtained at different sites.

The data reported in table V.1 on low-noise road pavements were obtained with this method. Naturally, to assess the global effect of the pavement on the roadside, it is necessary also to make the measurements with the "traffic" source, which generates the "horn" effect. As a result, a combined survey method is currently under study to measure the rolling noise, by means of a multi-wheel trailer, as well as on-site noise absorption.

In Austria, studies and research are currently under way on further test methods, such as acoustic impedance or absorption. In the Netherlands, efforts are being devoted to developing a representative method for measuring and evaluating the acoustic properties of new types of tyres.

V.3. DESIGN, CONSTRUCTION AND MANAGEMENT OF LOW-NOISE ROAD PAVEMENTS

The effective noise abatement of certain road pavements depends not only on the reference surface used for the measurements but also on the following factors:

- sound source considered;
- test vehicle (type and speed);
- tyre type;
- road geometry (alignment, elevated or depressed slope of road);
- weather conditions.

Pavement noise performance depends upon the pavement structure type and mix characteristics and their evolution over time:

- surface distress;
- clogging process;
- deterioration of structure;
- maintenance schemes.

For use as a reference surface, a special ISO Working Group has suggested a dense, smooth-textured asphaltic concrete surface with maximum chipping sizes of 11-14 mm. But it is questionable whether this reference surface, as made in different countries, would yield acoustic performance with an acceptably low variability.

Before examining the influence of the intrinsic and technological characteristics of mixes on pavement traffic noise abatement, summarising the references in this regard in the OECD countries and in the bibliography, certain observations are appropriate regarding mix design and placement, in view of what was stated in the first section concerning the influence of their surface characteristics on sound emissions.

V.3.1. Influence of surface characteristics

It is possible to act on the texture of a pavement, and consequently on the noise it emits, through careful choice of the sizes and shape of the aggregates to be used in the surface layer mix. According to the studies and applications concerning this point, the sizes of aggregates used for this purpose should not exceed 8/10 mm. One should add that it is preferable to choose maximum sizes on the order of 4-6 mm, thus limiting this characteristic of the aggregate, rather than try to obtain high amplitude wave to contain the noise emitted at high frequencies. But the sizes and types of aggregates should also be chosen with the polishing effect of traffic taken into account. Certain crushed aggregates facilitate this, even in the case of microtexture surfaces (i.e. textures with wavelengths of 0.1-0.5 mm).

Pavement wear generally leads to an increase in the high frequencies, due to the polishing process or the clogging of draining asphalts. If the integrity of the surface is altered, there is an increase in low-frequency noise emissions.

To ensure the homogeneity of the road surface, we must use aggregates of uniform type and nature, and roll them in such a way as to ensure a non-random orientation and suitable surface features.

Low-noise surfaces of types A, B, C, and D perform acceptably with regard to skid resistance.

Excessive pavement stiffness, which results from a high-stiffness binder, seems to have an effect on noise. Here, it is not yet clear what effect is obtained by adding rubber granules to the asphalt.

Above, in the first section, it was stated why road pavements with porous surfaces are able to reduce traffic noise emissions. It is, in fact, true that pavements with an optimal surface texture can perform well in terms of tyre-road noise, but cannot achieve the performance levels attainable with acoustically optimised porous surfaces.

Porosity not only reduces air pumping and resonances, and the acoustic amplification effect ("horn" effect), but also contributes significantly to acoustic absorption, influencing multiple reflections and noise propagation (excess attenuation).

Table V.3 shows the various porous asphalt mixes used in OECD countries. It is generally considered that draining asphalt with a maximum chipping size of 10-12 mm and an air voids content of 22-23 per cent is the most effective.

Sound absorption by draining asphalt depends on conservation of the porosity. Porosity is maintained through washing by traffic rolling on wet roads. Moreover, it has been proven that the greater the permeability of the drainage asphalt after placement, the longer good performance will last. Also, reducing the sand content of the mixture significantly improves the hydraulic properties, the limitation being the ability to maintain the integrity of the surface (aggregate stripping or local loss of material) and mechanical strength under traffic. It seems that the voids content must be limited to 25 per cent (interconnecting voids > 20 per cent).

With regard to the sound absorption coefficient, optimisation will lead to high values for frequencies in the range of 500-1 500 Hz on high-speed roads, whereas on urban roads it is preferable that such values be attained at lower frequencies (250-1 500 Hz). It should be recalled, furthermore, that the band width is influenced by:

Table V.3.a. **Characterising data of porous asphalt -- Routine use**

Country	Layer thickness mm	Voids content % vol	Mineral aggregates				Binder			Reason for use		Start of res.		Quantity (total amount in 1992) x 1000 m²	Remarks
			Max size mm	Stone content % mass	Gap mm	Hydr. lime % mass	Pen. grade mm 10^{-1}	Content % mass	Conv/Modif	Safety	Noise	1st	2nd		
A	40	≥22	11	82	2/8			4.8-6.3	mod.	•	•	84		7500	Maintenance problems to be solved
B, L	40	22	14	83	2/7	≤2	80/100	4-5	conv/plasto/ elasto	• exc.L	• exc.L	79 exc.L			
								5.5-6.5	recycled elasto						
E	40	20	10	86	0.63/2.5		60/70	≥ 4.5	80 % mod.	•	•	80			Doubts on durability
			12.5	84	2.5/5		80/100						86		
F	30-40	≥ 20	10	88	2/6		60/70 40/50 80/100	4.5-5.5	mod.	•	•	77		21.000 (15.000 motorways)	Doubts on durability and winter maintenance
			14	82	2/10					•			84		
I	40	≥ 18	14	88	2/7		80/100	5.0-6.5	mod.	•	•	87		10.000 (motorways)	Sound absorption is an important pavement performance characteristic
NL	50	≥ 20	11	85	2/6	1.3-2.3		4.5	conv.	•		72			Winter maintenance problems in 78/79
			16					4.5-5.5	mod. (exp)		•		85		

Table V.3.b. **Characterising data of porous asphalt -- Experimental use**

Country	Layer thickness mm	Voids content % vol	Mineral aggregates				Binder			Reason for use		Start of res.		Quantity (total amount in 1992) x 1000 m²	Remarks
			Max size mm	Stone content % mass	Gap mm	Hydr. lime % mass	Pen. grade mm 10^{-1}	Content % mass	Conv/Modif	Safety	Noise	1st	2nd		
D	30-60	≥ 20	8 11	85	2/5		80	≥ 5.0	mod.		•	79	86		Experimental stages
DK	40	≥ 16	8	78			100	≥ 4	conv.	•		73			Stopped in 77/79: winter maintenance problems
			12	82							•		90		Experimental stages
UK	50	20	20	90	3/6	2	200 100 70	3.7 5.0	conv. mod.			< 70	84		Durability problems
GR			19	82	2/6		80/100	5.0	mod.			> 86			Initial investigations
P		≥ 20	15					5.0-5.5	mod.	•	•	90			Initial investigations
Summary of typical properties	**40/50**	**≥ 16**	**10-14**	**85**	**2/7**		**≤ 100**	**≥ 4.5**	**Polymer modified**	•	•				

- thickness;
- porosity and tortuousness;
- air flow resistance.

The best experimental results have been obtained on high-speed roads with draining asphalt 6-8 cm thick (F, DK, NL: twinlay). For urban streets, greater thicknesses could lead to better results, but using draining asphalt in such cases is not recommended because of the sharp drop in performance due to clogging (two years on average S, F).

V.3.2. Experience in some OECD countries

In **Australia,** investigations have focused on the most widely used types of surface layers, such as hot treatments with aggregates having maximum sizes up to 14 mm, cold bituminous microcarpets, porous asphalt, and rigid and traditional asphalt pavements. Porous asphalts have been studied mainly to compare their performance when the binder used is varied. They proved to have lower noise level values, especially those constructed following revision of the 1987 specifications. It is interesting to note that in the 10 dB(A) over the maximum shown by the porous asphalts there was no sharp division between the types in concrete and those in asphalt.

In **Austria**, the use of low-noise pavements, including not just porous asphalt but also microcarpets, surface treatments of small maximum size, and concretes with exposed aggregate, has led to recognition of the need to optimise the maximum size, the amplitude, and the wavelengths of the surface texture.

Porous asphalts, which in Austria as elsewhere have proven the most effective among the aforesaid noise abatement techniques, have according to the relevant Austrian standard a maximum aggregate size of 11 mm. Earlier test sections had shown that best performance in terms of sound absorption was at a maximum aggregate size of 8 mm, but these mixtures led to serious stability problems and shorter durability when subjected to high volumes of freight traffic.

Concerning rolling noise, the measurements made in Austria give the values 1.5, 3.5, and 5 dB(A) as reductions in permanent equivalent sound energy with respect to traditional asphalt surfaces at speeds of 50, 60, and 80 km/h, even though the spectrum of single measurements reached as much as 7-9 dB(A), despite the fluctuations resulting from the inhomogeneity of the only comparisons between porous surfaces it has been possible to make up to now. Austrian tests on porous asphalt showed substantial and rapid reductions in traffic noise, especially after the first winter, and in correspondence with the wheel paths, despite difficulties experienced in cleaning this type of pavement.

The Austrians have used microcarpets (<25 mm) and surface treatments based on epoxy resins. The former, which make it possible to do without lateral drainage devices, have been used on urban roads, with aggregates 4 to 8 mm in maximum size, yielding notable noise reductions at high frequencies, whereas the latter have been used on pre-existing rigid pavements, on both urban and extra-urban roads, with the epoxy resins used as is or blended with cement and aggregates of 3-4 mm maximum size.

Another way to achieve substantial noise reduction with respect to other concrete pavements is to apply a concrete surface course 3-4 cm thick with aggregates of very small maximum size (8 mm, with the largest possible proportion of 4/8 aggregates). The day after construction, the fine mortar on the top and between the aggregates is brushed off ("exposed aggregate concrete"). The surface corresponding to Type A of figure V.4 will lead to a maximum reduction of rolling noise.

In **Belgium,** research on the long-term behaviour of porous asphalts, i.e. monitoring of 3 experimental sections for 52 months, found the noise level on the 20-mm-thick section to be twice that recorded on the 40-mm sections.

In **Denmark,** particular attention has been devoted to porous asphalts, in particular to monitoring reference sections and experimental sections on both urban roads and motorways to determine reductions in sound levels, as well as the effect of the filling of residual voids over time. The experiments conducted in urban areas with aggregates having maximum sizes between 8 mm and 12 mm and residual voids of 20-23 per cent and 24 per cent showed greater noise reduction in only one case. With the more limited number of voids, the noise reduction with respect to the traditional mix proved to be less, but this, as already explained, limited its decay over time, especially with aggregates of 12 mm maximum size and at low frequencies.

In **Finland,** the construction of pavements with surface layers selected for noise abatement is a rather recent development, and studies have centered on comparing the performance of porous asphalt surfacing with that of traditional asphalts and rigid pavements and with surface treatments using large aggregates. The measurements, conditioned by the presence of studded tyres, have shown that in the case of bitumen modified with rubber granules the sound spectrum is shifted toward that of concrete surfaces. Measurements made before and after the winter season showed an increase in the noise emission level.

In **France,** interesting research has been conducted to explain the reduction in sound emissions by porous asphalt pavements, studying the air pumping and "horn" effects, as well as to determine the manner of spatial propagation of noise according to the type of tyre. Researchers have also studied an algorithm designed to describe and predict "horn" phenomena using a calculation model. Several experiments were carried out to check different hypotheses regarding optimised porous surfaces. The parameters investigated included thickness (from 3 to 50 cm), single and multiple layers, mixture types, aggregate types, binders, additives, etc., on sections located on both high speed roads and urban streets. There are currently 21 million m^2 of draining asphalts, placed mainly on motorways (15 million m^2). The use of draining asphalts on urban streets has been given up after unsuccessful experiments in Paris, Lyon, and Nantes. The current trend is to lay high-void-content draining asphalt only on high-speed roads, and in certain areas with a very limited percentage of sand, giving high levels of permeability and, consequently, increasing the durability of this fundamental performance characteristic of draining asphalt. For urban streets, experiments under way show that thin and ultrathin microtexture layers of bituminous concrete are effective. The expected reduction using draining asphalt as compared to 0/14 mm bituminous concrete is at least 3-4 dB(A) in long-term LEQ.

In **Germany,** studies have been conducted on types of macrotexture, and the clearest results are the discovery that the maximum acoustic benefit, in terms of rolling noise abatement, is obtained with pavements made with aggregates having a maximum size of 11 mm. The experiments have confirmed that the acoustic absorption curves obtained with the standing wave tube (Kundt) on pavements differing solely with respect to the type of binder are practically identical. Furthermore, it was found that an increase in pavement thickness shifted the peak resonance in the acoustic absorption curve toward the low frequencies. The global reduction of traffic noise emissions in terms of dB(A) is practically independent of the thickness of the pavement in the 1-3 cm interval.

Moreover, there was found to be a strict dependence between the energy absorbed and the percentage of residual voids in the mix, but even a level of 25 per cent was not sufficient to produce a reduction of more than 2 dB(A) in an arbitrary noise produced radially in the pavement. This research used vehicle type approval procedures to study the rolling noise emissions of an automobile passing with engine shut off, at constant speed, on two different types of pavement, one a porous asphalt surface with high maximum aggregate size and the other a traditional pavement with a smooth surface. The results indicate a lower spectrum energy content on the part of the porous asphalt for frequencies above 1000 Hz, whereas below this frequency the noise produced by tyre vibrations on the road predominates. In this case there is an increase in rolling noise on draining pavements.

Also in Germany, experimental sections were monitored for 34 months to determine rolling noise using the trailer method. As a function of the maximum sizes of the aggregates, there was found to be a decline in rolling noise over time for the values of 11 and 16 mm, whereas for the porous asphalt with maximum sizes of 3, 5, and 8 mm, there is an increase in the noise emitted, although originally these gave the lowest values. This is explained by what was stated above in section V.2 regarding the trailer measurement method.

In **Japan,** a country which has for years been committed to traffic noise abatement, there has been a movement for some time now to favour construction of porous asphalt surfaces so as to limit the high costs entailed by noise abatement barriers. The experiments conducted to date have confirmed that, with increasing vehicle speeds and increasing residual voids, there is an increase in noise abatement in comparison with traditional pavements, but this is not true of the thickness. Increasing the level of residual voids beyond 20 per cent tends to slow the rapid decay of noise abatement, thanks to the use of modified bitumens.

Future developments are expected to reduce the costs of porous asphalt in comparison with those of traditional mixes, by means of special long-term contracts with construction firms and the adoption of particular specifications.

In **Italy,** research aimed at rolling noise abatement through optimisation of road pavements was begun started in 1987, simultaneously with the construction of the first porous asphalt pavement. Today, draining and sound absorbing pavements cover more than 10 million m2 of Italy's motorways. An attempt has been made to take a methodological approach to the problem, first of all through separate analysis of the various mechanisms and phenomena that play a role in the production of rolling noise as perceived at the roadside, and secondly through the development of reliable experimental techniques for measuring the acoustical properties of pavements, both in the laboratory and on site (for this latter case see section V.2). The third phase, currently at the stage of real large-scale verification, is aimed at optimising those characteristics proven to have the most influence on rolling noise.

In the first stage, on-road measurements were taken to characterise noise sources, while at the same time assessing the existing measurement methods used for this purpose; in the second stage, a special methodology was developed to measure the sound absorption coefficients of porous asphalt draining and sound-absorbing pavements. Subsequently, after the effectiveness of these pavements was evaluated in semi-anechoic chamber tests, an innovative sound-absorbing pavement was designed and subsequently subjected to scale-model testing in the laboratory.

In **Norway,** in 1988, several experimental road sections were built to check at full scale the behaviour of certain types of road surfaces formulated to obtain a reduction in traffic noise emissions. For porous asphalt pavements, in this country too, emphasis was on aggregate gradings; with a maximum size of 11 mm, it was found that the optimal residual voids percentage from the standpoint of long-term conservation of acoustic properties was 22-23 per cent. This, combined with greater care in cleaning the aggregate, would have avoided the "husking", and consequent material loss, which took place during winter operations, together with slipperiness phenomena, in the initial experiments, which in any case gave very promising results from the acoustic standpoint when rubber granules were used as aggregate.

The most recent approach makes it possible, even without effective maintenance systems, to conserve the main surface characteristics of these pavements by optimising mix design, production, and placement. One of the results found was that after initial decay during the first summer season, these pavements held up well against the wear caused by studded tyres. These are used on a large percentage of vehicles during the winter season in Norway, and cause rapid breakdown of pavements that are not well designed. Simultaneous experiments with different methods of cleaning these draining pavements revealed the impossibility, because of the damage caused by the studded tyres, of restoring the sound absorption to its original levels; this is in contrast with the positive results obtained with respect to water removal. Porous asphalt is not recommended in Norway for this reason.

Similar research on optimising the surface characteristics of existing rigid pavements for noise abatement purposes is still in the study stage.

In **The Netherlands,** the most recent experiments on low-noise pavements had to do in particular with porous asphalt, initially introduced in the road sector for safety reasons. The application technique employed with these pavements entailed the laying of two distinct layers, one 45 mm thick consisting mainly of 11/16 mm aggregate, and a 25 mm surfacing with 4/8 mm aggregate. These draining pavements have residual voids of 26 per cent when opened to traffic, and the binder is a modified asphalt with rubber granules, 0.15/1 mm in size, equivalent to 16 per cent of the binder weight.

The limited thickness of the surface layer of this draining pavement, adopted in the most recent experiments (1989-91) conducted for sound absorption purposes, made possible, with respect to the first applications (1986), especially on urban roads, an increase in noise abatement in comparison to traditional mixes [4 dB(A) and 5 dB(A) for 60 and 120 km/h, respectively]. Further abatement can certainly be achieved with the appropriate choice of tyres.

In **Spain,** porous asphalt is frequently used for anti-noise surface courses. These pavements have been used since the late 1980s on roads of all categories, and now cover 30 million m2. The mixes are normally laid in a 40-mm thickness with a maximum aggregate size of 12 mm, 10-15 per cent undersize on a 2.5-mm screen, and a 4.5 per cent filler content, with over 20 per cent residual voids.

Binders are generally 4.5 per cent of the weight of aggregates; modified binders are used in 70-75 per cent of cases (mostly for motorways and main roads); in the remaining 25-30 per cent of cases normal 60/70 dmm penetration bitumen is employed. A new porous asphalt mix design method based on the Cantabro test has been developed: Marshall specimens are subjected to weight loss by rolling using L.A. equipment without the steel balls normally used for tests on aggregates. It has been found, after two years of service, that porous asphalts having residual voids lower than 20 per cent tend to fill in, but retain good levels of skid resistance.

Again in Spain, porous concrete pavements able to limit noise emissions by vehicles have been developed within the framework of a European Community programme carried out with Danish and German participation.

Apart from confirmation of results concerning these low-noise pavements, already checked in other past experiments, it is worth recalling the efforts spent studying the correlations between tyre-road noise, surface texture, aquaplaning, and skid resistance in both applications. Based on the first results, which seem to show the absence of any reliable correlation between tyre-road noise and texture depth, as measured using the sand patch test, we are trying to define another texture index.

In **Sweden,** tyre-road noise was surveyed using the following three techniques for on-site measurement of pavement acoustical properties: "coast-by", "drums coated" and "trailer" methods.

The "coast-by" is a measurement described by type-approval rules for tyred vehicles, and in this case it is applied by test vehicles travelling with engines off and in neutral. The "drums coated" is carried out in a silent chamber: different tyres are tested on benches equipped with rolling drums, and noise emission at different speeds is measured. The "trailer" method measures tyre-road noise emission on the road.

Among other things, these measurements have identified possible operations to limit vehicle noise emission, expected to result in a 10 dB(A) decrease in the equivalent overall level in twenty years, thus minimising the effects of the expected growth of the number of road vehicles. In addition to pavements, these operations should concern vehicles and tyres as well as road users, through a suitable road regulation system.

Concerning porous asphalts, one Swedish experiment should be recalled: using mixes with 12 mm max. aggregate sizes, a very wide initial residual voids range (20-30%), and three different thickness values, it revealed that over a four-year period noise reduction decreases with time, perhaps because of the effects of the cold seasons, and that the average reduction over four years is 3 dB(A), with a larger reduction for the equivalent level measured instead of average peak levels for all vehicles, independently of the type of vehicle. Moreover, this difference is the same as that recorded between new and old dense bituminous pavements.

In addition, it was found that as thickness is increased, noise reduction increases as well, although this effect decreases over time, and initial residual voids should be at least 20 per cent, compatible with mix durability requirements.

In **Switzerland,** experiments have been conducted in several Cantons to investigate the traffic noise abatement of certain road pavements, with special attention to those types of surfacings with high residual voids content. Currently under way are experiments aimed at optimising the relevant parameters in order to identify the characteristics of a wider range of road surfacings suitable for noise abatement purposes.

In the case of porous asphalt pavements, this research was also pursued with the aim of improving the useful life of such pavements; the initial experiments showed positive results from the standpoint of sound absorption, but suffered premature breakdown of the surfacing as compared to traditional pavements.

In the **United Kingdom,** where porous asphalts were introduced for the first time, especially in airport pavements, results are available from the long-term monitoring of their acoustic behaviour. These results, which appear encouraging, at least for certain definite conditions, show a noise reduction of 3-6 dB(A) in the first year, with higher levels for light automobiles and lower levels for freight vehicles, and 4 dB(A) for the next 4 years, falling to 3 dB(A) at 5-6 years (the same level for both light and heavy vehicles).

In the **United States,** no particular priority appears to be given to the study of tyre-pavement interaction from the acoustic standpoint, mainly because of the ambiguous and especially non-lasting influence of pavement type and texture on noise. For these reasons, the technical-economic choice of type of road surface will depend mainly on other considerations; note also the stated difficulty of forecasting the future condition of the upper pavement layers. Moreover, the aforesaid heterogeneity and equivocal nature of the results of the type of pavement on noise emission control does not permit consideration of this factor in the analyses, which are nevertheless being conducted on road traffic noise in general. Nevertheless, tests on porous asphalt pavements constructed with bitumen modified with rubber granules have confirmed the sound absorption reported in European experiments.

Finally, in the United States, stone mastic asphalts are reported to show less noise abatement than draining pavements, even though their influence is specific to the higher frequencies (1000-5000 Hz).

V.3.3. Management and maintenance of porous asphalt

At this point, in view of the fact that porous asphalt is the mix most widely used in producing sound absorbing pavements, it is in order to provide a brief summary as to how, by adopting a series of management measures summed up in figure V.8, the useful life of porous asphalt draining and sound-absorbing pavements can be optimised. These management rules naturally involve the conservation of the drainage and sound absorption functions of the intercommunicating voids of the mixes. The first function should be designed in accordance with pre-determined rainfall intensities (medium-high), and also according to the width of transverse drainage (in the case of straights) or the crossfall (in the case of curves), so as to ensure, over time, the skid resistance necessary for driving safety. To increase water removal capacities, one could also act on the thickness, but this is done, for obvious reasons, only at certain points on the road, for example, on straight-curve transitions, where there is insufficient crossfall.

The internal drainage of rainwater can be reduced and impeded by the following conditions:

- Non-uniform porosity in the transverse direction, because of the way the course is laid, in several contiguous strips (sealing of joint zone). This lack of uniformity may also be caused by differentiated compaction due to heavy traffic on the strips traversed most frequently by the wheel paths of freight vehicles;
- Progressive filling of the sides of the pavement through the accumulation of occluding materials, or water transport (from inside to outside), or the fall of material from roadside earth embankments;
- Original lack of suitable aggregate sizes or of adequate intercommunication of voids resulting from the grading.

For more details about construction specifications, the reader is advised to consult the many published reports available (for example, the PIARC report on porous asphalt).

To maintain the draining function over time, it is sometimes necessary to perform preventive (better) or remedial unclogging operations. The common methods are high-pressure water cleaning or a suck-sweep cleaning truck (a special machine, which washes the pavement with hot water, applying pressure to one part and a vacuum to the other part, so washing out fine or soluble filling materials with the liquid). These washing actions are performed periodically (by way of indication, based on Italian

Figure V.8. **Management cycle of porous asphalt**

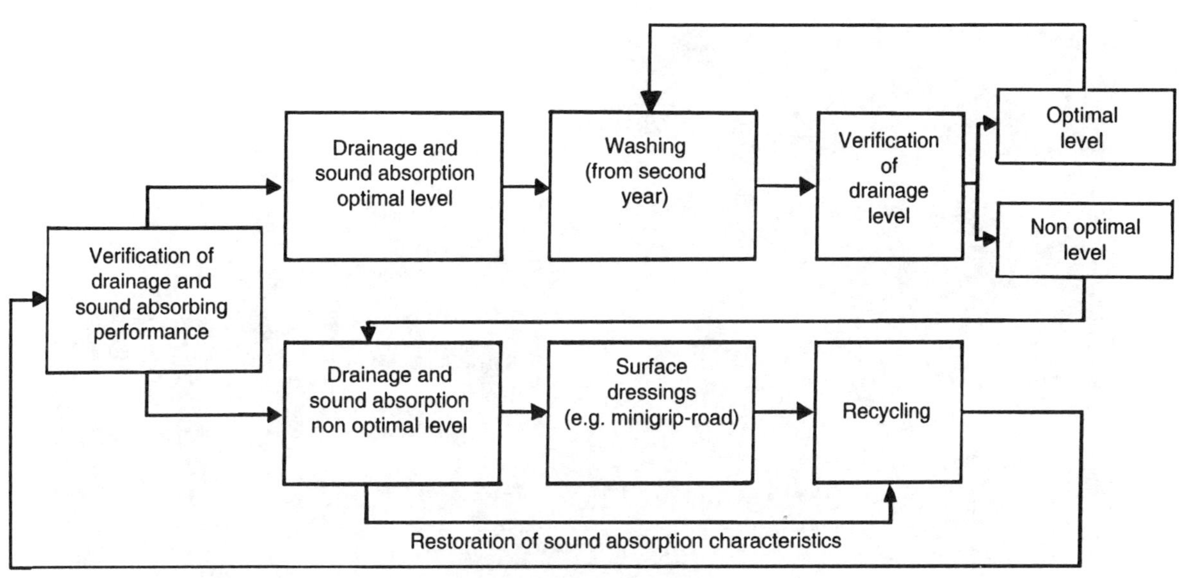

experience, every two years), starting from the first or second year of operation, preferably on the shoulders and, in the case of motorways, on the slow lane; it is also advisable to perform washing operations on shoulders whenever maintenance work on embankments may have reduced the perviousness of this zone, before the penetrating materials can become consolidated.

Naturally, if the lack of porosity in certain zones is related to the limited percentage of residual voids or the presence of longitudinal "sealing", the washing action will be ineffective and useless. The validity of the operation can be checked by measuring the hydraulic conductivity before and after the work.

Accumulation of dirt over long periods can also render washing operations useless, since the filling material becomes cemented and impossible to remove even by means of the powerful cleaning actions described above (this generally occurs about 3 years after initial construction).

When draining asphalt becomes clogged, the skid resistance and surface texture values must be checked. If there is total drainability, the surface may have reduced skid resistance and texture values (especially for purposes of tyre/road noise abatement). However, if there is insufficient drainage, the skid resistance and texture values must be improved to ensure tyre/aggregate contact even through a film of water.

For these reasons, therefore, it may be necessary to perform surface roughening operations:

- Mechanical type, to increase the micro-roughness of the aggregate (with recovery of the abrasive so as not to fill the remaining voids);

Figure V.9. **"Hurricane" declogging machine**

- Overlay type, to be performed using micro grip road, which increases the texture without completely compromising the sound-absorbing function.

This latter impairs the perfect continuity of the internal voids, insofar as sound energy is much more "penetrating" or, in other words, less viscous than water; the size of the voids, however, influences the range of sound frequencies absorbed. There is also an effect due to the smoothness of the surfaces of the small channels within the mix, the "dirtying" of which can increase sound absorption while impairing the draining function.

In any case, it can be held that the sound absorbing function is preserved longer than the draining function, sometimes even with a slight positive variation followed by a deterioration in frequency characteristics.

The other characteristic of porous mixes, sound absorption, is relevant to tyre-road noise, linked to pavement surface texture. Where, when porous asphalt was first used, there was a tendency to give great value to this type of noise abatement and to use "flat" aggregates to reduce noise generation, the trend today, as already mentioned, is to optimise sound absorption due to the "horn" effect, without reducing the microroughness of the aggregates or the macroroughness of the surface, so as to compensate for possible loss of the drainage function over time, thus making the surface potentially more slippery when wet. Simultaneously with this tendency, there is a move to optimise the quantity of voids in porous asphalts, using ever more "open" graded mixes.

Maintenance of the draining function does not alter the noise abatement function in any clear-cut way. Only after the roughening treatment are there certain reductions or shifts in the frequencies absorbed, but these are always acceptable by comparison with the traditional pavement.

As stated above, at the end of the useful life of the draining function, a roughening operation is in order; this is true if one is content to forego the advantage of reduced water spray and if the noise abatement function is either still satisfactory or else unnecessary (porous asphalt surface layer far from any urban zones). Otherwise, the draining and sound absorbing functions must be reconstituted, and unsuitable materials or laying choices adopted during construction possibly corrected. In such a case, one can act by recycling, using a hot milling machine, possibly correcting the original grading curve with the addition of aggregates; some are particularly recommended, such as pre-bitumened foamed clay. Experiments of this type, as already mentioned, have been conducted in France, the Netherlands, and recently also in Italy.

Even though porous asphalt pavements may have maximum drainability, they can pose management problems in winter, because sometimes de-icing salts can be leached into the pavement or otherwise not remain on the surface long enough to eliminate possible glare ice or frozen sleet. For this reason, larger quantities of de-icing salt are used on these pavements, or else special "viscous" salts obtained by mixing NaCl and CaCl2, which remain longer on the surface and within the mix.

Failure to take these last measures will not have objectively risky consequences, but may cause ice to form within the porous asphalt, which can give rise to the fear of a "potentially slippery pavement", especially for drivers of freight vehicles, who may halt, slowing traffic. These precautions can be avoided if the above indications are followed, or if preventive de-icing salt treatments are applied in a timely manner.

V.4. REFERENCES

1. PIARC, Technical Committee on Surface Characteristics (1979). *Etude du bruit résultant de l'interaction des pneumatiques et de la route.* Report to the XVIth P.I.A.R.C. World Road Congress, Vienna, September, 1979.

2. PIARC, Technical Committee on Surface Characteristics (1983). *Bruit de contact pneumatique-chaussée.* Report to the XVIIth PIARC World Road Congress, Sydney, September, 1983.

3. OECD ROAD TRANSPORT RESEARCH (1984). *Road Surface Characteristics: Their Interaction and their Optimisation.* OECD, Paris.

4. PIARC. Technical Committee on Surface Characteristics (1987). *Road Traffic Noise and Surface Characteristics Optimization.* Report to the XVIIIth PIARC World Road Congress, P.I.A.R.C. reference 18.01.E, Brussels, September, 1987.

5. DESCORNET, G. (1988). *Surface Characteristics Optimisation Criteria.* Proceedings of International Road and Traffic Conference on "Road and Traffic 2000", vol. 2, pp. 329-333, Berlin, September 6-9, 1988.

6. DELANNE, Y. (1989). *Porous Asphalt Acoustical Optimisation.* Proceedings of the International Symposium on "Aménagements routiers et sécurité", Luxembourg, June 14-17, 1989.

7. T.R.B. (1990). *Porous Asphalt Pavements: An International Perspective 1990.* Proceedings of the T.R.B./P.I.A.R.C. Joint Session on "Porous Asphalt Performance", Transportation Research Record, N. 1265, Washington, 1990.

8. CAMOMILLA, G., MALGARINI, M. and R. GERVASIO (1990). *Remarks on the Performance of an Extensive Network of Sound-Absorbing Porous Asphalt Pavement.* Proceedings of the Annual Meeting of the Transportation Review Board, 1990.

9. BÉRENGIER, M. and J.F. HAMET (1990). *Porous Asphalt Acoustical Properties: Sound Absorption Phenomena.* Bulletin de Liaison des Laboratoires des Ponts et des Chaussées, vol. 168, July-August 1990. LCPC, Paris.

10. SPRINT. (1991). *Bituminous Road Surfacing.* Proceedings of the First S.P.R.I.N.T. Workshop, Exhibition and Demonstrations on "Technology Transfer and Innovation in Road Construction", Lisbon, April 22-24, 1991.

11. PIARC, Technical Committee on Surface Characteristics (1991). *The Interaction Between Road and Motor Vehicle.* Report to the XIXth P.I.A.R.C. World Road Congress, P.I.A.R.C. reference 19.01.B, Marrakesh, September, 1991.

12. DESCORNET, G. (1990). *Reference Road Surfaces for Vehicle Testing?* Roads/Routes P.I.A.R.C. Magazine, n. 272,III, 1990

13. IVF/LCPC/CETUR (1992). *The Mitigation of Traffic Noise in Urban Areas.* Proceeding of the Eurosymposium, May 12-15, 1992.

14. SANDBERG, U. (1992). Low Noise Road Surfaces-A State-of-the-Art Review. Proceedings of Eurosymposium on "The Mitigation of the Traffic Noise in Urban Areas", Nantes, May 12-15, 1992. LCPC. Paris.

15. PIPIEN, G. (1992). *Chaussées peu bruyantes - Problématique générale.* Proceedings of Eurosymposium on "The Mitigation of the Traffic Noise in Urban Areas", Nantes, May 12-15, 1992. LCPC. Paris.

16. VAN MEIER, A. (1992). *Thin Porous Surface Layers - Design Principles and Results Obtained.* Proceedings of Eurosymposium on "The Mitigation of the Traffic Noise in Urban Areas", Nantes, May 12-15, 1992. LCPC. Paris.

17. T.U. Berlin/PIARC/TRB/ASTM/FGSV. (1992). Proceedings of the Second International Symposium on Road Surface Characteristics, Berlin, June 23-26, 1992.

18. CAMOMILLA, G., M. MALGARINI and R. GERVASIO (1992). *Optimization of Sound Absorption Performances of Porous Asphalt Pavements.* Proceedings of the Second International Symposium on Road Surface Characteristics, Berlin, June 23-26, 1992.

19. IRF/FRI (1992). Contribution towards Reduction of Traffic Noise. International Road Federation Working Group on "Interaction of Vehicles, Tyres and Pavement", Paris, March 1992.

20. PIARC (1992). *Noise reducing concrete surfaces.* Proceedings of a PIARC-Workshop held 24-25 February 1992 in Vienna, Schriftenreihe Strassenforschung, BMwA, Wien, 1992.

21. SOMMER, H. (1992). *Noise Reducing Concrete Surfaces/State-of-the-Art.* Proceedings of a PIARC Workshop held on 24-25 February 1992 in Vienna, Roads-Routes Magazine, n. 278, vol II, 1992. PIARC.

22. SANDBERG, U. (1992). *Low Noise Road Surfaces-Design Guidelines.* Proceedings of the Second International Symposium on Road Surface Characteristics, Berlin, June 23-26, 1992.

23. PIARC, Technical Committees on Flexible Pavements and Surface Characteristics (1993). *Porous Asphalt - Les enrobés drainants.* PIARC reference 08.01.B, 1993.

24. SANDBERG, U. (1993). *Measuring Method for Comparing Noise on Different Road Surfaces.* Status Report ISO/TC43/SC1/WG33. 22 February, 1993.

25. DELANNE, Y. and D. SOULAGE (1993). *Etat des connaissances sur les couches de roulement peu bruyantes.* November 1993.

26. AUTOSTRADE SpA (1991). *Pavement Maintenance: Performance Tender Technical Specifications.* Autostrade Company, Research and Maintenance Central Directorate, January 1991, Rome.

27. AUTOSTRADE SpA (1992). *Instructions for Environmental Impact of Road Structures Regarding Acoustical Pollution Control.* Internal document of Autostrade Company, Research and Maintenance Central Directorate, prepared for the National Instructions for Environmental Impact of Road and Railway Structures. Rome.

28. AUTOSTRADE SpA (1993). *Input for the Pavement Maintenance Via Electronic Programming* (M.A.P.P.E. Project): Pavement Program 1994. Internal report of the Autostrade Company, Research and Maintenance Central Directorate, December 1993, Rome.

29. SOMMER, H. (1993). *Lärmmindernde Betonoberflächen (low noise concrete surfaces).* Schriftenreihe Straßenforschung, Heft 415, BMwA, Vienna.

30. LITZKA, J. and W. PRACHERSTORFER (1994). *Österreichische Erfahrungen mit Lärmmindernden Straßendecken (Austrian experiences on low noise pavements).* Schriftenreihe Straßenforschung, Heft 427, BMwA, Vienna.

31. PIARC (1995). *Reports to the XX th World Congress.* Montreal, September 1995. PIARC.

CHAPTER VI

NOISE BARRIERS

VI.1. INTRODUCTION

The basic physical principles for noise reduction must be applied to the design process of an acoustically effective noise barrier, i.e. a barrier that provides the required noise reduction without being "overdesigned". Non-acoustical features of noise barriers, such as maintenance, safety, aesthetics, construction, cost, etc. must also be considered in noise barrier design. The main design criteria for noise barriers are summarised in table VI.1. The importance of community (public) participation in final design and implementation decisions is stressed.

For a noise barrier to work, it must be high and long enough to prevent the propagation of noise toward the receiver. For example, the normal noise barriers do very little good for homes on a hillside overlooking a roadway or for buildings which rise above the barrier. Openings in barriers for driveway connections or intersecting streets destroy their effectiveness. In some areas, homes are scattered too far apart to permit noise barriers to be built at a reasonable cost. The acoustic performance of a noise barrier is determined by its location, length and height, and transmissive and reflective (insulating)/absorptive characteristics.

Noise barriers can be constructed as:

- Natural screens (barriers) like earthberms;
- Artificial screens like walls;
- Mixed types like biowalls or screens on earthberms.

VI.2. ACOUSTICAL CONSIDERATIONS

Roadway traffic noise can be attenuated by the construction of noise barriers between the roadway and areas adjacent to the roadway. A noise barrier substantially interferes with the propagation of the sound waves from the roadway to the receiver. The sound waves are bent or diffracted over the top of the barrier, creating a "shadow zone" (see figure VI.1) behind the barrier where noise levels are lowered. Effective noise barriers can reduce noise levels by 10 to as much as 25 decibels [dB(A)]. A 10 decibel reduction cuts the loudness of traffic noise in half.

Table VI.1. **Main design criteria for noise barriers**

Acoustical considerations:

Noise attenuation qualities are of foremost interest i.e. the acoustic efficiency of barriers in terms of sound frequencies and intensities abated. Key points are:

- A noise barrier can usually reduce noise levels by 10-15 dB(A) in the shadow zone;
- A noise barrier can achieve a 5 dB(A) noise level reduction when it is tall enough to break the line-of-sight from the roadway to the receiver;
- A noise barrier can achieve an approximate 1.5 dB(A) additional noise level reduction for each metre of height after it breaks the line-of-sight (with a maximum theoretical total reduction of 20 dB(A);
- The length of a noise barrier should be approximately 4 times the distance from the receiver to the barrier; and
- A noise barrier should have a minimum density of 20 kg/m^2.

Non-acoustical considerations:

Noise barrier design should include appropriate consideration of:

- Aesthetics; especially in regard to the surrounding environment;
- Traffic safety, i.e. in terms of drivers visibility and resistance to vehicle impacts;
- Maintenance (and maintenance costs); i.e. of the barrier itself and the surrounding structures, as well as drainage requirements;
- Structural performance (wind and traffic actions, long-term stability) and life expectancy of generally 15-20 years; and
- Construction costs depending on the type of foundation necessary and the construction method under traffic or not.

VI.2.1. Principles and mechanisms

Figures VI.1 and VI.2 present the key acoustical principles of relevance to noise barrier design and location. Noise reduction measured at the reception point R due to the barrier is called "insertion loss" (see figure VI.1). This is due to the extension of the path of the diffracted sound wave (see also figure III.3) to the barrier induced insulation in the shadow zone and the absorption.

Figure VI.1. **"Insertion loss" due to erection of noise barrier**

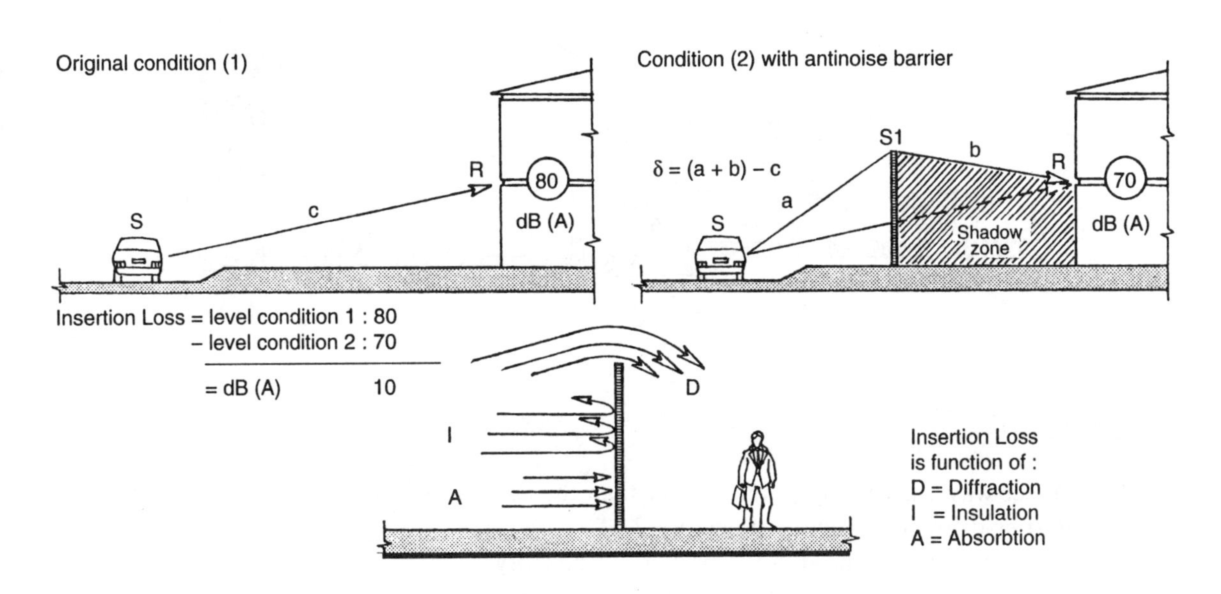

It is also worth recalling that by interposing a screen between a source S and a reception point B, sound propagation is heavily altered, and several elementary sound paths can be distinguished (see figure VI.2). To optimise the noise screen, each of the contributions should be minimised, since they are the most significant in terms of noise pollution (except for effect "6" which is to be maximised). It can be seen that it is impossible to define an optimal barrier in absolute terms. However, it is correct to look for a suitable barrier for each particular condition.

VI.2.2. Implications for planning and design

The height and location of a noise barrier relative to the roadway are important acoustical considerations in its design. At a fixed distance from the roadway, increasing the height of the barrier will increase its attenuation characteristics. A value of 1.5 dB(A) per incremental metre of height may be used to approximate the height of barrier needed to achieve a desired attenuation, assuming that a barrier which just breaks the line-of-sight (straight line from the roadway source to the receiver) provides a 5 dB(A) attenuation.

For a constant barrier height, moving the barrier close to the receiver or close to the source increases its attenuation. However, in practical design, it may be possible to take advantage of local terrain conditions to find a barrier location which can benefit from higher elevations. Figure VI.3

Figure VI.2. **Propagation of highway noise in the presence of a noise barrier**

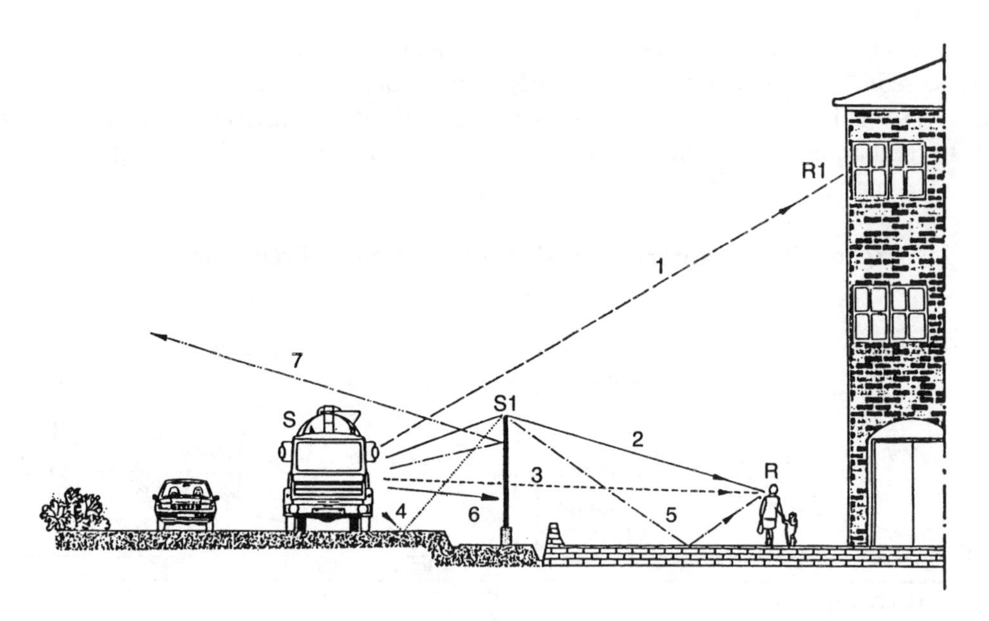

Notes:

1. Direct wave, involving points higher than the SA line -- trajectory 1
2. Barrier-diffracted wave -- trajectory 2
3. Wave transmitted through the barrier -- trajectory 3
4. Ground-reflected wave, subsequently diffracted -- trajectory 4
5. Diffracted wave, subsequently reflected by the ground -- trajectory 5
6. Absorbed wave, for sound-absorbing barriers
7. Barrier-reflected wave -- trajectory 7
8. Wave that, after multiple reflections between barrier and source, travels over the top of the barrier

illustrates that a shorter barrier placed on hilly terrain provides more attenuation than a higher, and therefore more expensive, barrier located closer to the roadway.

Roadway noise can travel around the end of a barrier and reach the receiver if the barrier is too short. A rule-of-thumb to ensure that the barrier is long enough to avoid this undesirable effect is that the barrier should extend four times as far in each direction as the distance from the receiver to the barrier.

VI.2.3. Implications for choice of materials

In addition to sound that travels over the top of the barrier to reach the receiver, sound can travel through the barrier itself. The amount of "transmission" through the barrier depends upon factors relating to the barrier material (such as its weight and stiffness), the angle of incidence of the sound, and the frequency spectrum of the sound. As a general rule, if the transmission loss through the barrier is at least 10 dB(A) above the attenuation resulting from the diffraction over the top of the barrier (or the barrier density is approximately 25 kg/m^2 minimum), the barrier noise reduction will not be

Figure VI.3. **Noise barrier placement**

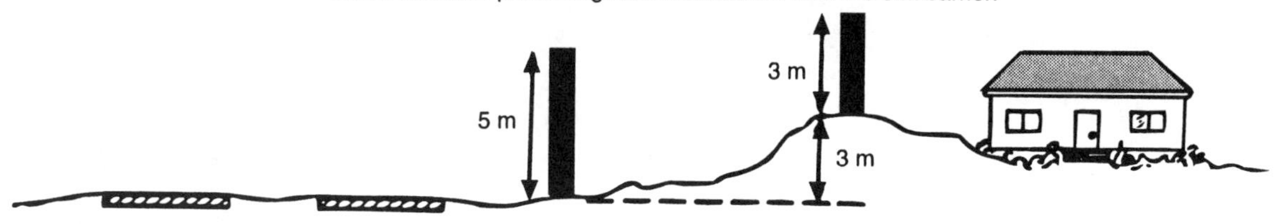

substantially affected by the transmission through the barrier. It should also be noted that barriers are more effective in abating higher frequencies of noise, since the smaller wavelengths associated with higher frequencies are more easily affected by solid objects such as barriers.

Sound energy is be reflected when a sound wave strikes a solid object, such as a noise barrier. Reflections by a single barrier wall to the opposite side of a roadway would be less than 3 dB(A), since this single reflection can at most double the sound energy. There has been concern that multiple reflections (see figure VI.4) of sound waves between two parallel plane surfaces, such as between noise barriers or retaining walls on both sides of a roadway or between noise barriers and the sides of passing trucks, when the number of trucks is great and the distance between the barriers and the trucks is small, can increase noise levels in the immediate area.

Austria has conducted studies which indicate an increase of more than 3 dB(A) in the instance of barrier-truck reflections and, as a result, uses only absorbing barriers where dwellings are located on the opposite side of a roadway. Other countries (e.g., the United States), where roadway cross-sections are larger and lateral distances between /parallel plane surfaces are greater, have not identified a problem with multiple reflections. Sound energy can also be absorbed when a sound wave strikes a solid object; the absorptive characteristics for any particular material are a function of the frequency of the sound.

Noise barriers may therefore be designed to be:

- *Insulating* (reflective), when they reflect the noise back to the side origin and prevent its transmission through the barrier, thus acoustically insulating the adjacent shadow zone;

- *Sound-absorbent,* when the sound wave is dampened (in terms of frequencies), due to reflections, interferences and other phenomena which take place within the barrier itself.

A sound-absorbent barrier will not have an insulating function if the constituent material is too thin or too light. In practice, this should however not occur. Well-made sound-absorbent barriers are thus insulating and reduce the multiple reflections discussed above. The materials that can be used are extremely varied. In table VI.3, some elements for barrier choice are given and the various types and materials used are described in later sections of this Chapter.

Figure VI.4. **Multiple noise reflections. Problems generated by parallel, non sound-absorbing barriers**

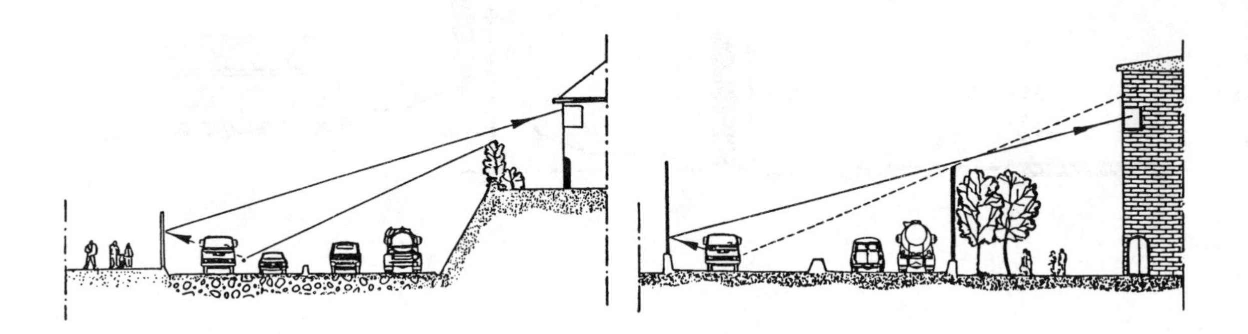

VI.3. AESTHETIC CONSIDERATIONS

VI.3.1. Visual effects

A major consideration in the design of a noise barrier is the visual impact on the adjoining land use. An important concern is the scale relationship between the barrier and activities along the roadway right-of-way. A tall barrier near a low-scale single family detached residential area could have a severe adverse visual effect. In addition, a tall barrier placed close to residences could create detrimental sun shadows and thus affect the microclimate. One solution to the potential problem of scale relationship is to provide a stepped noise barrier to reduce the visual impact through introduction of landscaping in the foreground. This allows additional sunlight and air movement in the residential area. In general, it is desirable to locate a noise barrier approximately four times its height from residences and to provide landscaping near the barrier to avoid visual dominance.

The visual character of noise barriers should be carefully considered in relationship to their environmental setting. The barriers should reflect the character of their surroundings as much as possible. Where strong architectural elements of adjoining activities occur in close proximity to barrier locations, a relationship of material, surface texture, and colour should be explored in the barrier design (for example, type 8 barriers - see table VI.3). In other areas, particularly those near roadway structures or other transport elements, it may be desirable that proposed noise barriers have a strong visual relationship, either physically or by design concept, to the roadway elements. In general, a successful design approach for noise barriers is to utilise a consistent colour and surface treatment, with landscaping elements used to soften foreground views of the barrier. It is usually desirable to avoid excessive detail which tends to increase the visual dominance of the barrier.

VI.3.2. Effects on drivers

The psychological effect on the passing motorist must be taken into consideration too. Barriers should be designed differently to fit dense, urban settings or more open suburban or rural areas and should also be designed to avoid monotony for the motorist. At normal roadway speeds, visual perception of noise barriers will tend to be of the overall form of the barrier and its color and surface texture. Due to the scale of barriers, a primary objective to achieve visually pleasing barriers is to avoid a tunnel effect through major variations in barrier form, material type, and surface treatment.

The design approach for noise barriers may vary considerably depending upon road design constraints. For example, the design problem both from an acoustic and visual standpoint is substantially different for a straight roadway alignment with narrow right-of-way and little change in vertical grades than for a roadway configuration with a large right-of-way and variations in horizontal and vertical alignments. In the former case, the roadway designer is limited in the options of visual design to minor differences in form, surface treatment, and landscaping. In the latter case, the designer has the opportunity to vary the barrier type, utilise landscaped berming, and employ more extensive approaches to develop a visually pleasing barrier.

VI.3.3. Barrier lay-out

From both a visual and a safety standpoint, noise barriers should not begin or end abruptly. A gradual transition from the ground plane to the desired barrier height can be achieved in several ways. One concept is to begin or terminate the barrier in an earth berm or mound. Other concepts include bending back and sloping the barrier, curving the barrier in a transition form, stepping the barrier down in height, and terminating the barrier in a vegetative planter. The concept of terminating the barrier in a vegetative planter should only be utilized in areas where climatic conditions are conducive to continued vegetative growth and in areas where the planter edges will be protected from potential conflict with roadway traffic.

VI.3.4. Graffiti

Graffiti on noise barriers can be a potential problem. A possible solution to this problem is the use of materials which can be readily washed or repainted. Landscaping and plantings near barriers can be used to discourage graffiti as well as to add visual quality.

VI.3.5. Summary

Aesthetic design and the integration of noise barriers into the landscape and the environment are of special importance. This is especially true of barrier height, the choice of material, and the barrier shape, structure, and colouring. A successful design approach for noise barriers should be multidisciplinary and should include architects/planners, landscape architects, roadway engineers, acoustical engineers, and structural engineers.

VI.4. OTHER NON-ACOUSTICAL CONSIDERATIONS

VI.4.1. Public involvement

Public involvement should be a vital part of good noise barrier planning and design. It is important to take into consideration the wishes and desires of the residents, since they must live with the barrier each day for many years into the future. The residents may be very unhappy and perceive no noise reduction benefits from a barrier they deem unsightly and had no part in planning. It is also desirable to involve officials and organisations in the affected area in planning for a barrier.

VI.4.2. Safety

It is desirable to locate a noise barrier out of the area where it may be struck by a vehicle which may stray from a roadway. Safety barriers may be needed to protect noise barriers at risky locations, particularly when noise barriers are constructed within the right-of-way of an existing roadway.

The safety barrier can be constructed in front of or as part of the noise barrier. When the available space is limited, the existing road safety barrier may be integrated in the noise barriers. This is the case with types (7) included in table VI.3, and the D and E foundations in table VI.2. These protections are also effective in preventing heavy vehicles straying from roads. Note that these designs have passed > 485 KJoule crash tests in compliance with the new standard set by C.E.N. (*Comité Européen de Normalisation*).

Consideration must also be given to safety when locating noise barriers in the vicinity of on- and off-ramps, ramp intersections, and intersecting roads. A noise barrier should not block the line-of-sight between the vehicle on the ramp and approaching vehicles on the major roadway.

Snow removal considerations become a safety factor when the melting snow forms ice on the roadway surface or when blowing and drifting snow accumulate on the trafficked lanes. Sufficient space must be left for snow ploughing. Noise barriers should also not shade the roadway surface in such a manner as to encourage ice formation. The surface treatment of the barrier also has safety implications. Protrusions on a barrier near a traffic lane, and facings which can become missiles in a crash situation, should be avoided.

VI.4.3. Maintenance

Maintenance considerations for noise barrier design include maintenance of the barrier itself; maintenance associated with adjoining landscaping; replacement of barrier materials damaged by vehicle impacts; and cleaning the barrier and/or removal of graffiti. In general, maintenance of barrier materials is less costly if materials with unpainted surfaces such as weathering steel, concrete, pressure-treated wood, or naturally weathered wood are used. It is desirable from a visual and maintenance standpoint to use concrete surfaces which are left natural, such as sandblasted finish and exposed aggregate, or with integral colour or coloured cement mortars, as opposed to painted surfaces which require continual long-term maintenance. Maintenance of landscaping associated with the edge of the road right-of-way will be affected by both the noise barrier placement and the type of landscaping used.

Providing access to the rear of the barrier for maintenance purposes or emergency access by varying the horizontal alignment of the barrier can also provide visual relief. In general, access breaks in the barrier should be designed to avoid an abrupt wall facing the flow of traffic. Where a solid door is not provided for access, overlap parallel barrier walls can be constructed, with a minimum of three times the width of the opening and absorptive facing, to maintain the acoustical effectiveness of the barrier (see figure VI.5).

Figure VI.5. **Overlap parallel barriers**

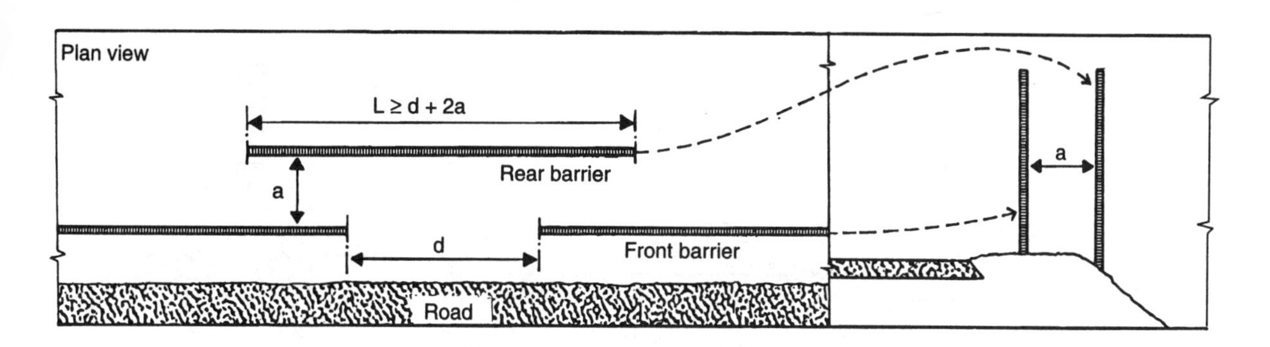

Another maintenance consideration with noise barriers is maintaining a stock of materials which are compatible with the barrier for replacement. This can be a serious problem, especially with naturally weathered finishes on steel and wood. Lastly, snow and ice removal may affect maintenance when earthberms are used as noise barriers. Planting materials used on the earthberm must be resistant to the effects of salt and other chemicals used in snow and ice removal. Snow removal costs may be increased when tall noise barriers are located very close to a roadway (limited space is available for pushing snow to the side of the roadway) and snow must be blown over the barriers or carried away by trucks.

VI.4.4. Drainage

Water drainage along the roadway right-of-way can be seriously affected by the construction of a noise barrier. Appropriate consideration for drainage should be made early in the planning stage of construction. Drainage structures should be designed to extend under noise barriers. Openings should not be made in the barriers themselves to provide for drainage, since, they can substantially degrade the acoustic performance of the barriers.

This potential loss of acoustic insulation is not taken into account with the same attention by all the countries. Some of them, such as The Netherlands, tolerate the presence of drainage openings under noise barriers to avoid higher costs for soundproof water drains. A possible solution is given in figure VI.6 and in line with the principle described in figure VI.5. The bent curb should be equal to or longer than twice its width (a), increased by the opening (d) visible from the road.

Appropriate consideration should also be given to the watering requirements of vegetation planted for landscaping around noise barriers. An artificial irrigation system may have to be provided in areas where natural water and drainage systems are insufficient to maintain vegetative growth.

VI.4.5. Barrier foundations

Foundation design for noise barriers should be based upon established theories and accepted testing and must mainly take into account:

Figure VI.6. **Plan view of a soundproof water drain (the curb walls should be porous)**

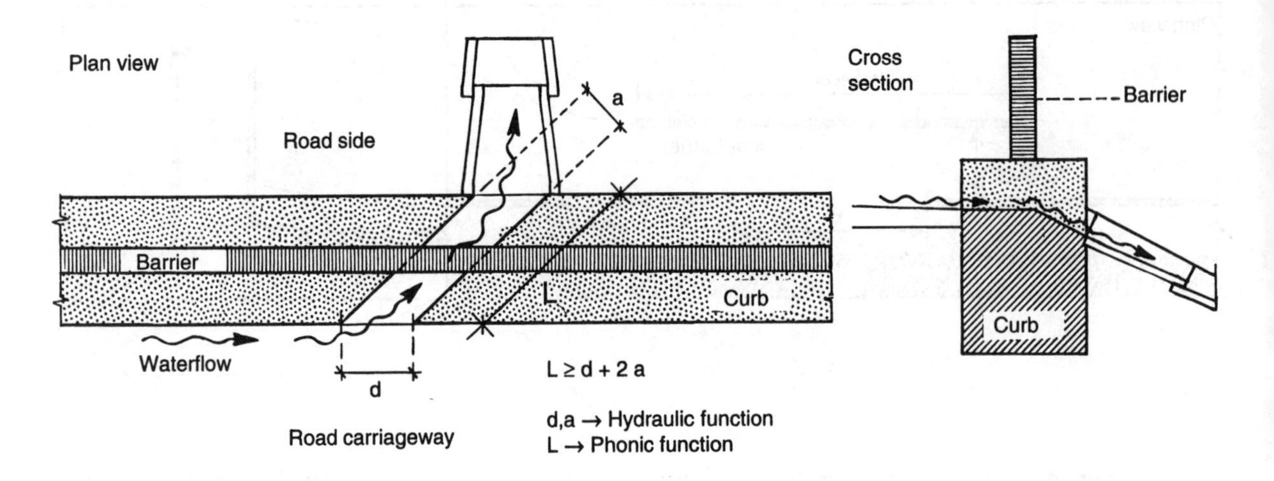

- *Wind effects* affecting structural stability due to the very large surface of noise barriers;

- *Differential settlements;* apart from the negative aesthetic effects, they may reduce noise-reduction performance of seals at the contact points between barrier elements, and

- *Costs:* foundations may have a considerable weight on total costs, particularly in very windy regions, and may account for as much as 30 per cent of the cost of the complete noise barrier.

Many countries have developed original solutions for these problems. For instance in Italy, curbs or plates are fixed to the support New Jersey barriers, these are moveable, when impacted, due to ductile anchor devices that "control" the elements which may be shifted up by 40 cm.

The United States has found another solution by water filling caisson holes for piles where groundwater is very close to the surface of the ground. The "auger-cast" pile method forces pressurized concrete from an auger with an internal passage into the hole as it is bored for the caisson. As the auger reaches the bottom of the hole, pumping of concrete begins, and the auger continues in rotation while being removed. In this way, concrete under pressure is pumped into the hole, preventing water from entering the hole. Precast posts with extended reinforcement bars are then placed in the newly poured concrete.

The most widespread systems are isolated plinths (B) and continuous curbs (D) with technical variations (C-type) (see table VI.2).

VI.5. TYPES OF NOISE BARRIERS

Different types of noise barriers are used in OECD Member countries. They have been classified by shape and conditions of use in table VI.3.

Table VI.2. **Types of foundations**

Type of foundation	Description of foundation	Type of barrier	Safety protection	Utilisation field	Relative cost
A 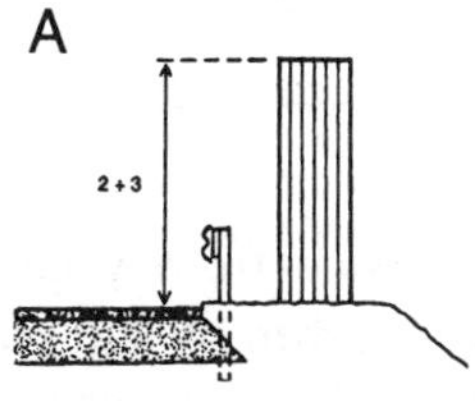 "No" foundation	A1 - Zigzag interconnected panels laid on earth l = f(h). A2 - Light panels with a small curb (30 x 30 cm).	A1 - Reinforced concrete (RC), heavy panels; max. h.: 2.50÷3.000 m. A2 - Light panels (wood, aluminium); height < 2.00 m. Integrating green walls.	Metal barrier (or earthmound)	Level ground, solid soil. Wide lateral space.	0.1 to 0.4
B 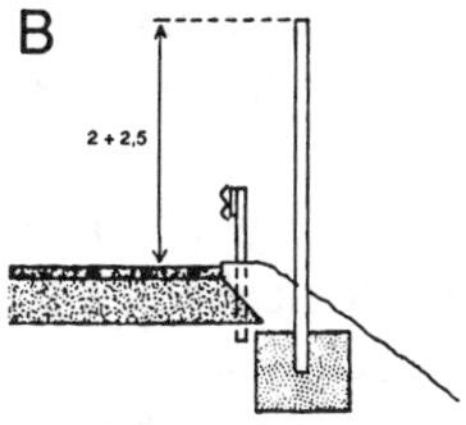 Foundation on isolated plinths	RC isolated plinths; typical distance: 3 m. Sized to stand local wind Barrier max. height: 2.00 ÷ 2.50 m.	Light panels connected to vertical mounts. Possibility to employ different materials in function of location.	Metal barrier.	Low embankments (h < 3.50 m). Level ground. Ground with a good bearing capacity	1
C 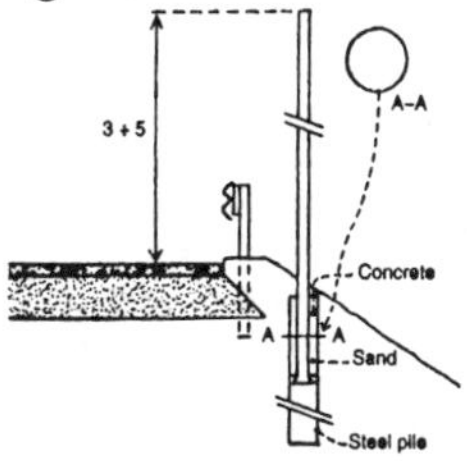 Foundation on steel piles	Steel piles stuck in the ground and mount fixed with sand and cement. Depht in function of height and local wind.	Barrier as for B foundation. Barrier height: > 3.00 m up to 4.50 ÷ 5.00 m.	Metal barrier.	High embankments (h > 3.00 m). Ground with a low bearing capacity. Problems due to strong wind.	2.0
D 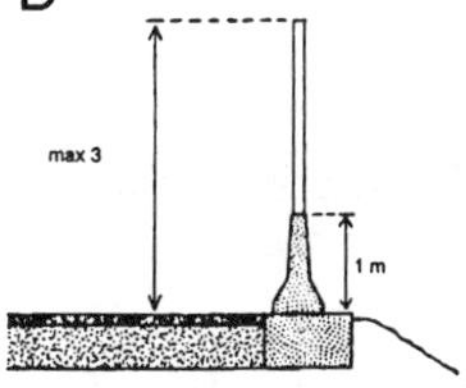 Foundation on curb	RC continuous curbs connected to NJ barrier by ductile anchor devices. Curb size function of wind (usually 0.60 ÷ 1.00 x 0.50 m).	Insulating and/or noise absorbing panels in RC even without mounts. Max. h: 3.00 m.	New Jersey self-protecting base (< 485 KJ crash energy tested).	High embankments (h > 3.00 m). Presence of things "to protect" out of road. Areas with medium-speed wind.	1.5
E 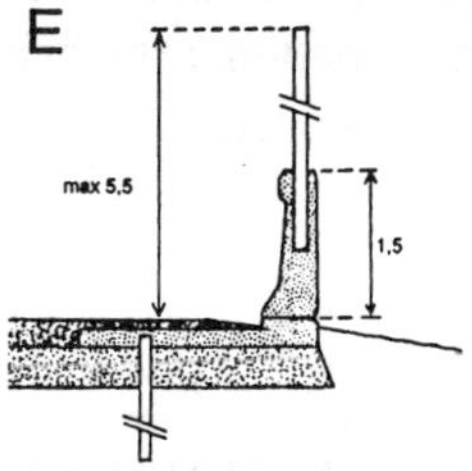 Foundation with nailed plates or curbs	RC plates (or curbs) connected to the ground with 12 micro-piles, vertical or tilted by 10° ("nails"), distance 1.50 ÷ 2.00 m Curbs are preferred for actions on already existing roads (1.00 x 0.60 m).	Barriers with light, very high (5.00 ÷ 5.50 m) panels laid on New Jersey foundations, also sound-absorbing for low frequencies.	New Jersey self-protecting base (> 485 KJ crash energy testec).	High embankments (h > 3.00 m). Approaching bridges with noise screens. For vey safe protection.	2.0

VI.5.1. Natural barriers

They are made of vegetation belts -- with a variable width of at least 10m - see figure VI.7 -- planted in a specified pattern. Plants are species that are selected in function of:

- ♦ height (grass, bush, shrub, plant)
- ♦ type of leaves (evergreen or deciduous)
- ♦ compatibility with climate (dry or wet regions)

Roadway noise levels can be reduced by means of absorption and diffusion from vegetation (diffusion increases the sound propagation area and part of the sound energy is absorbed by soil effects, air absorption, or friction with leaves, or is eliminated by conversion into heat). However, vegetation must be very tall and very dense to obtain a physical noise reduction. Vegetation that is planted as part of a roadway project provides especially a psychological noise benefit. It also has a visual psychological effect on local residents by shielding them from the constant sight of moving vehicles.

Noise reductions of up to 3 dB(A) have been obtained by Austria with natural vegetation (without special planting) 50-100 metres deep (smaller depths have produced only a psychological benefit). Italy has obtained noise reductions of 4-8 dB(A) with vegetation, depending on the species, height, intensity and position of the plantings. These results have been obtained by combining trees and bushes, planted in rows 6-7 metres deep, parallel to the roadway.

Natural barriers are often planted on suitable *earthberms* (figure VI.9). In general terms, *earthberms*:

- ♦ are typically covered with vegetation,
- ♦ have a very natural appearance,
- ♦ are usually attractive,
- ♦ typically allow more sunshine (less shade) and better air circulation than walls,
- ♦ can be used as a means of disposing of excess earthen material,
- ♦ normally do not require safety protection (guardrails) for vehicles that leave the roadway,
- ♦ are usually less costly to install and maintain than walls, and
- ♦ offer a practically unlimited useful life.

However, construction of earthberms can require extensive areas of land and considerable utility relocations.

Mixed barriers are obtained by inserting artificial screens -- made from wood or other material -- into natural barriers or over the top of earthberms, thus largely improving their noise-reduction performance. Mixed barriers also include artificial structures designed to allow the growth of grass, bush, shrubs, climbing plants or other vegetation hiding them to obtain the pleasant effect of natural barriers, sometimes in a limited space. These barriers are concrete, steel or wooden prefabricated structures containing a large volume of soil and often equipped with irrigation systems.

VI.5.2. Artificial barriers

Artificials barriers are the typical noise reduction barriers, and they are perceived as such even by non-experts. The way they work in terms of micro and macro acoustic has bee mentioned above. Artificial barriers may be classified as in table VI.3. There are:

Figure VI.7. **Natural barrier - Vegetation barrier with low buildings (approx. 10 metres)**

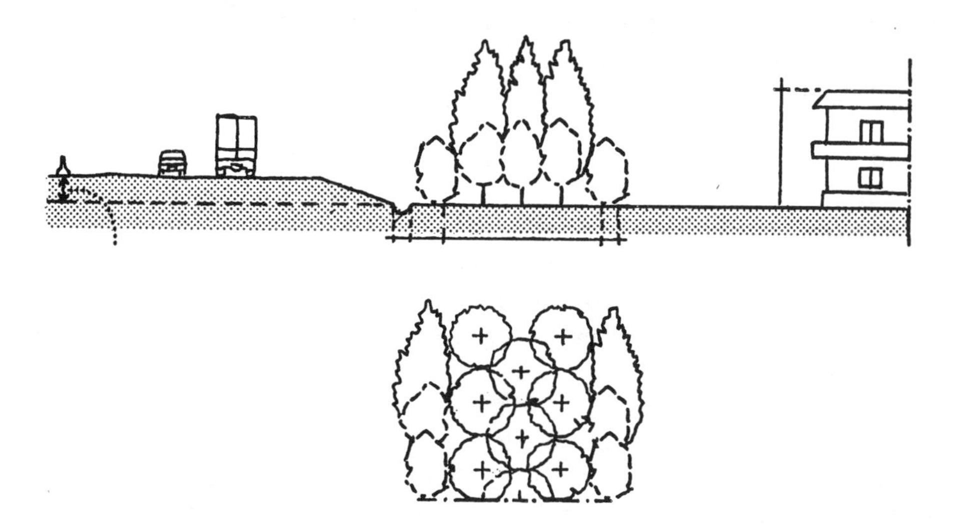

Figure VI.8. **Natural barrier - Earthberm (with vegetation) with buildings (10-12 metres)**

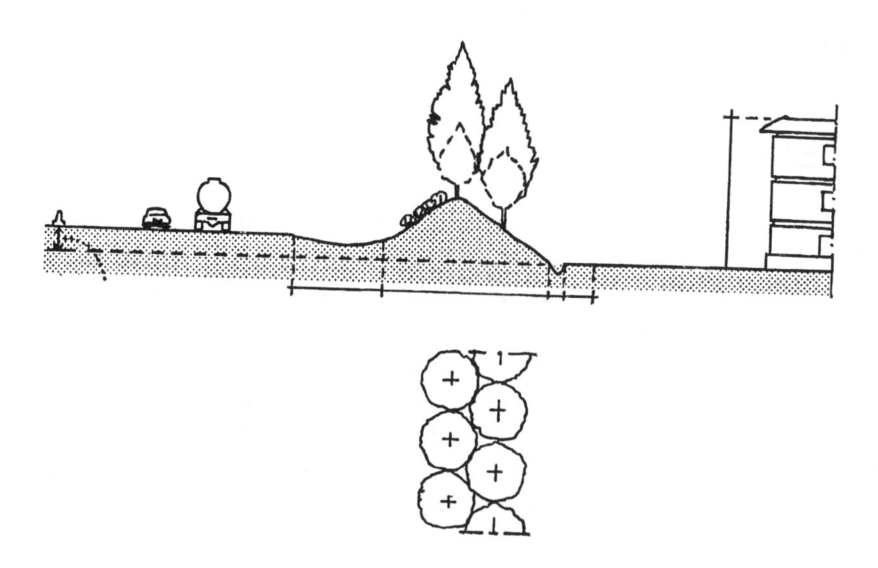

Figure VI.9. **Natural barrier - Low earthberm with biowall**

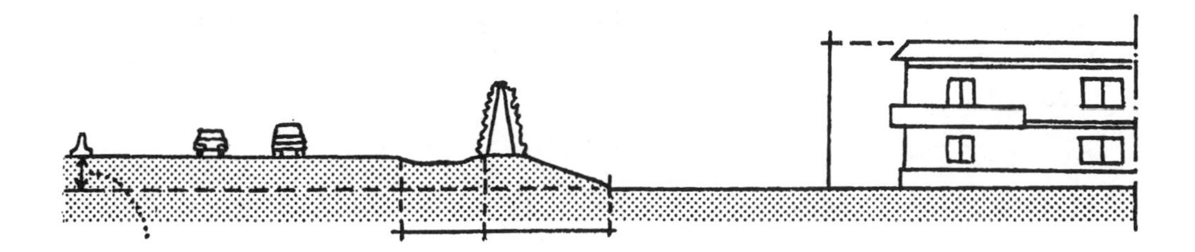

- continuous-structure barriers
- discontinuous-structure barriers
- continuous/discontinuous-structure, self-protecting barriers
- architectural barriers
- "total" barriers

These definitions do not take into account the materials they are made from, considering solely their geometry and morphology.

Continuous-structure barriers are made of elements connected continuously with one another. They provide good acoustic performance, but on the other hand they tend to instil a sense of oppression for travellers. However, their installation does not require much space, so they are suitable for being used in central urban areas.

The most widely used materials are concrete, wood and metal (aluminium or galvanized, painted steel sandwich, containing absorptive-fiber fillers), although some countries have a larger utilisation of wood or also of metal. Artificial barriers are often integrated by vegetation (bushes or climbing plants: excellent examples can be seen in Japan). In addition to improving the barrier from an aesthetical viewpoint, plants are also useful as they absorb emissions, particulates and heavy metals produced by traffic. For instance, there are "lead-eating" plants as *ailanthus glandulosus*; nevertheless, attention has to be paid to prevent their scattering by suitable pruning; plants could become noise generators over barriers, thus reducing their effectiveness.

VI.5.3. Assessment approach

In table VI.4, an assessment system in use in Italy for the classification of noise barriers is given. The same underlying scientific criteria are applied in France, Spain, Germany and in other countries (see also Chapter VII). Besides the type of material used and their intrinsic acoustical properties, the field tests necessary before any barrier can be approved -- as required by the established standards -- are indicated. They concern tests in regard to various sound sources, the average effects of insertion loss and the results of impulse tests. General technico-economic assessments are subsequently conducted, according to standard measurement conditions and in order to produce a data profile which can be compared to those of other barriers. As an example, data on a discontinuous aluminium barrier are provided together with a photograph (type 7 in table VI.3).

VI.5.4. Costs

Costs vary depending on materials and the kind of barrier using a specified category of material. Another cost sensitive element is the installation of barriers along existing roads, i.e. under traffic, particularly if nuisance to traffic has to be limited (night shifts, low-production worksites in limited spaces). This explains the variations shown in table VI.5 illustrating the cost range for barriers in OECD countries.

Table VI.3. **Noise barriers (based on Italian scheme)**

Types	Constituent materials	Dimension L=min length necessary H=min height	Acoustic functioning (**) I=insulation P=Phonoabsorbent	Area of application	Effectiveness (*) M; G; O; IL = dB(A)	+: Advantage -: Disadvantage
NATURAL BARRIERS						
1. Shrub Barriers	• Plants, grass • Bushes • Trees	L = at least 10m H = 8-9 m	P	Climatically suitable areas with sufficient space and receivers at low height 7-10m.	M IL = 3-4	+: pleasant aspect, also the absorption of exhaust gases. -:limited acoustic effectiveness, variable with the seasons
2. Earth Bermings with shrubbery	• Embankments & vegetation • same materials as above	L = 15-18 m H = 3-4 m	I-P	as in 1) but with wider spaces. Receptors with heights 7-10m.	G IL = 15-16	+: as in 1, but with higher acoustic effectiveness - needs ample space
MIXED BARRIERS						
3. Earth bermings with shrubbery and integrated artificial barriers	• As in 2, with barrier in wood, aluminium or other material	L = 15-18 m H = 9-12 m	I-P (I is higher with respect to 2)	As in 2 Ex: fig VI.11	O IL = 18-20	+: as in 2 with improvements -: as in 2
4. Biowalls or walls with vegetation	• Concrete or wood or metal materials containing vegetation & plants	L = 2-3 m H = 3-4 m	I (P = moderately)	As in 1, with little space available with receptor buildings nearby. Possibility of artificial integration. Ex.: fig. VI.18, 19, 24	G IL = 16-18	+: attractive aspect in limited space -: difficulties in plant growth Need for irrigation

ARTIFICIAL BARRIERS						
5. Continuous straight-line barriers	a in hollow brick or prefabricated elements b Panels in reinforced concrete c Panels in wood with or without absorbent material d Panels in aluminium or steel with absorbent material e In various materials f In transparent materials (metacrylic, polycarbonate)	a. L = 0.5 m H = 2.5 m b. L = 0.35 m H = 3-4 m c. L = 0.30 m H = 2-3 m d. L = 0.3 m H = 4-5 m e. l = 0.5 m H = 3-4 m f. L = 0.5 m H = 3-4 m	I & P (with hollows remaining) I & P I & P P I & P I	Roads which need severe noise abatement Limited availability of lateral space Urban & industrial environments Ex.= fig. VI.12, 14, 15, 25, 26, 2	a. G IL=15-16 b. G-O IL=17-19 c. G IL=18-19 d. O IL=20-22 e. O IL=15-19 f. G IL=16-17	+: good acoustic efficiency of the protection. Also reduced risk of fire and spin-offs -: problems due to the visual impact for road users and for drivers on the hard shoulders Increased difficulties for maintaining the hard shoulder
6. Discontinuous barriers	a Reinforced concrete panels b Panels in aluminium or steel with absorbent material c Mixed materials a or b with transparent panels	a. L = 1-2 m H = 3-4 m b. L = 1.0 m H = 3-4 m c. L = 2.0 m H = 3-4 m	I & P P I & P	Roads which need severe noise abatement Where lateral space available is average The need to see the panorama beyond the road Ex: Fig. VI.10	a. O IL=17-18 b. O IL=18-19 c. O IL=20-22	+: as in 5. Many of the contrast of 5 are reduced -: with respect to 5, require more space and higher maintenance costs
7.(***) Continuous & discontinuous straight-line barriers with anti impact basements	a Panels in reinforced concrete b Panels in aluminium or steel or wood c Mixed a or b with transparent panels	as in 5 & 6		As in 5 & 6 but with a safety basement/barrier for use with bridges or on road or for road borders which must not be crossed Ex: Fig. VI.17, 20	as in 5 & 6	+: as in 5 + protection for drivers -: as in 5
8. Architectural barriers	Composite barriers with designs, forms & artistic colours, various materials	L = variable from 0.5m to a few metres H = variable	I, rarely P	Harmonisation with environment if comprising: • natural attractions: does not detract from them • degraded ambience: improves or hides it Ex: Fig. VI.13, 15, 29	G IL=14-16	+: it is in keeping with the environment -: higher costs/maintenance costs
9. Total barriers	Type 5c or 5a barriers with sound-absorbing baffles in plastic material above the road	L = along all the road H = > 4.5m	P	Roads that need major noise abatement Where multi-storey buildings over 15m high need protection Ex.: fig. VII.3, VI.28	O IL= 20-25 5-8 floors at high of buildings	+: good protection even in high storey buildings -: as in 5 but worse

(*) EFFECTIVENESS M = Moderate, G = Good, O = optimum, IL = average insertion loss measured in some representative installations (10m behind the barrier, 1.5m height)
(**) Insulating is synonymous of "REFLECTIVE" ; Phonoabsorbent is synonymous of "ABSORPTIVE"
(***) Alternative solution is security steel barrier

Table VI.4. **Classification form for artificial barrier (Italy) (the examples are indicating data for discontinuous aluminium barrier)**

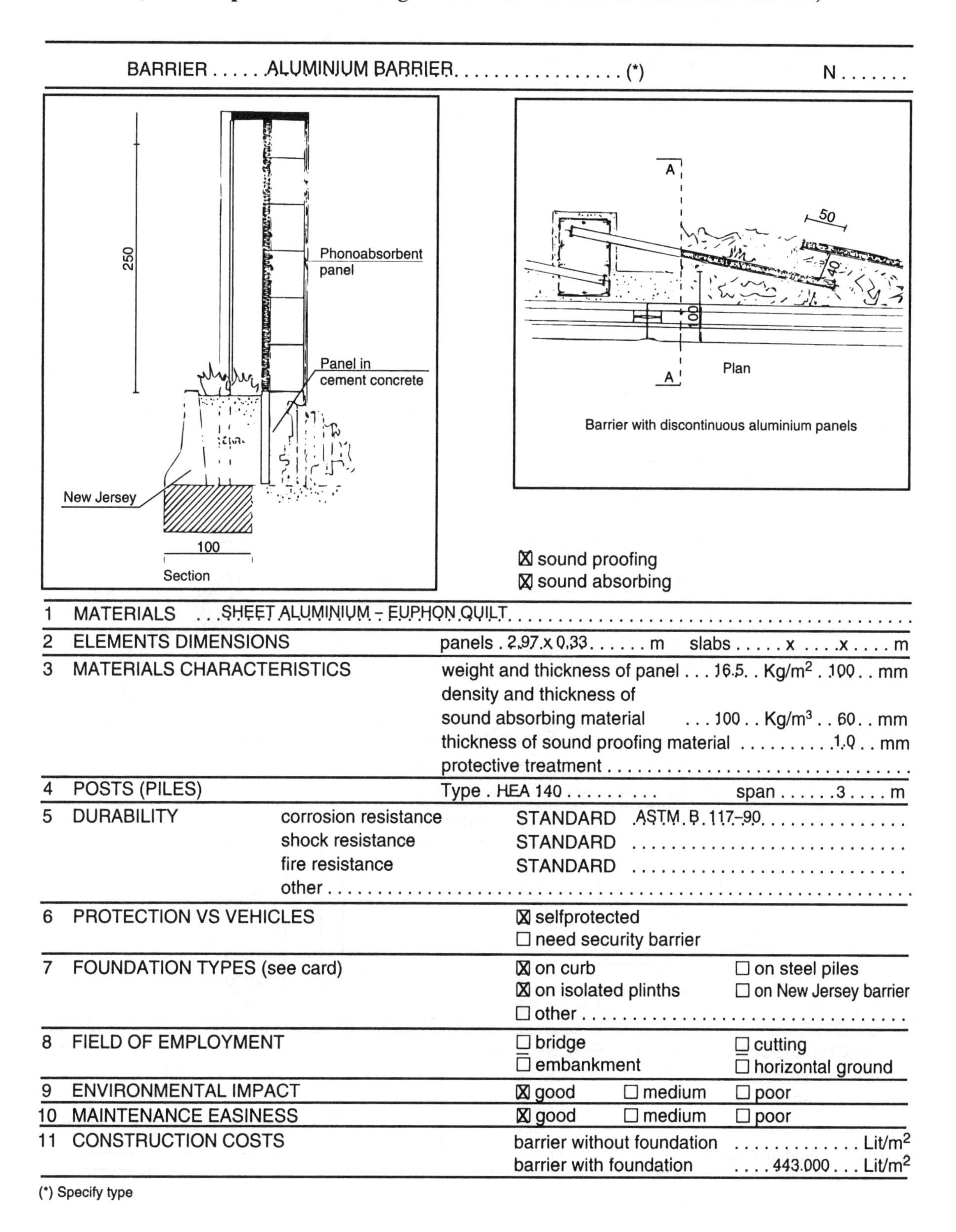

BARRIER ..... ALUMINIUM BARRIER .... (*) N ....

Barrier with discontinuous aluminium panels

☒ sound proofing
☒ sound absorbing

1	MATERIALS	...SHEET ALUMINIUM – EUPHON QUILT	
2	ELEMENTS DIMENSIONS	panels . 2,97 x 0,33 ...... m	slabs ..... x ....x .... m
3	MATERIALS CHARACTERISTICS	weight and thickness of panel ... 16.5. . Kg/m² . 100 . . mm density and thickness of sound absorbing material ... 100 . . Kg/m³ . . 60 . . mm thickness of sound proofing material .... 1,0 . . mm protective treatment	
4	POSTS (PILES)	Type . HEA 140	span ...... 3 .... m
5	DURABILITY	corrosion resistance shock resistance fire resistance other	STANDARD .ASTM . B . 117-90 .... STANDARD .... STANDARD ....
6	PROTECTION VS VEHICLES	☒ selfprotected ☐ need security barrier	
7	FOUNDATION TYPES (see card)	☒ on curb ☒ on isolated plinths ☐ other	☐ on steel piles ☐ on New Jersey barrier
8	FIELD OF EMPLOYMENT	☐ bridge ☐ embankment	☐ cutting ☐ horizontal ground
9	ENVIRONMENTAL IMPACT	☒ good ☐ medium	☐ poor
10	MAINTENANCE EASINESS	☒ good ☐ medium	☐ poor
11	CONSTRUCTION COSTS	barrier without foundation barrier with foundation	.... Lit/m² .... 443.000 ... Lit/m²

(*) Specify type

Table VI.4. **(following)**

BARRIER ..... ALUMINIUM BARRIER................. N.......

12 ACOUSTICAL PROPERTIES

Sound insulated and absorption test into reverberant room

surface of the material
for the sound absorption test ..11.47.m²
weight for surface unit ....Kg/m²
volume of the reverberant room ..205.9. m³
total surface of the room ..219.5. .m²
surface of the material
for the sound insulated test 7.7.m²
volume of the source and receiver rooms.52m³ .72m³
reverberation time at 1000 Hz ...... sec
diffusing surfaces ...... m²
air temperature .. 26. °C humidity .... 55. %
test sound . WHITE NOISE AT 1/3 OCTAVE...........
testing laboratory ....

Sound insulated ISO 717/1

R(dB): 0, 10, 20, 30, 40, 50, 60

125 250 500 1000 2000 4000 Hz

evaluation index .. 34 . dB(A)

Sound absorption ISO 354

α: 0, 0.2, 0.4, 0.6, 0.8, 1.0, 1.2

125 250 500 1000 2000 4000 Hz

125	250	500	1000	2000	4000
0.39	1.05	1.03	1.07	0.90	0.75

Other mesauring method ....
....
....

Field test

f (Hz)	white noise d1	white noise d2 (25)	pink noise d1	pink noise d2	traffic noise d1	traffic noise d2
125		7.5		11.2		7.5
250		6.8		11.0		15.0
500		5.8		19.0		14.
1000		13.3		18.0		24.0
2000		27.0		29.2		33.5
4000		17.7		22.5		31.0
		25.0		23.5		15.8

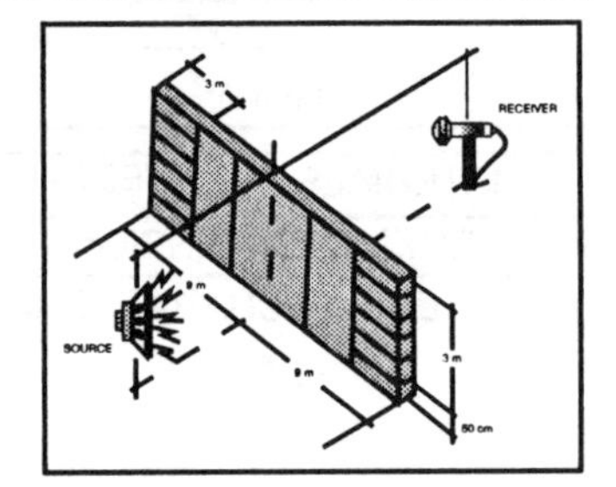

Insertion loss test (Δ Lx)

hx \ dx	d1 (10)	d2	d3
h1 (1.5)	24.3	24.5	25.8
h2	26.1		
h3			

Δ Lx = noise variation in dB(A) after the barrier insertion, measured at dx distance from the barrier and at hx height.

Impulsive test

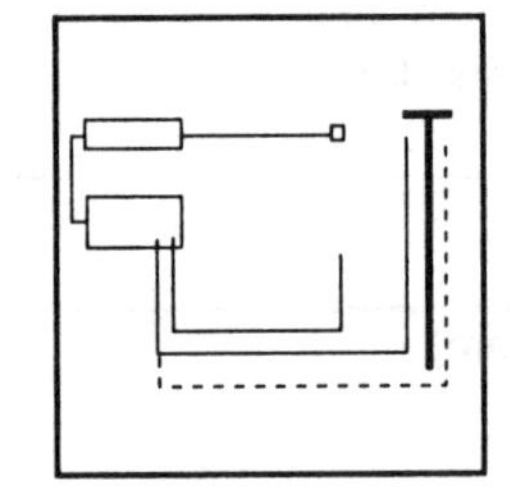

f (Hz)	α	R
125		
250		
500		
1000		
2000		
4000		

13 TECNICO-ECONOMIC GLOBAL EVALUATION

Δ L standard condition ...12.8. .dB(A)
Lit/m² (*) for dB(A) taken down
a) without foundation ...... Lit/dB(A)
b) with indicated foundation . 34.000 .Lit/dB(A)

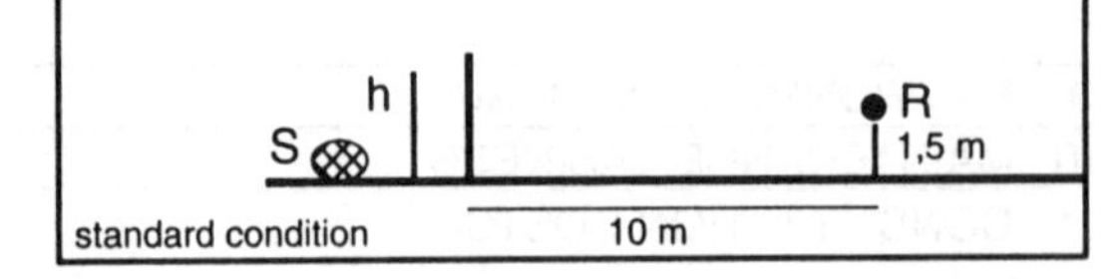

(*) see field 11

Figure VI.10. **Aluminium barrier -- absorptive "discontinuous system" -- as described in Table VI.4 (Italy) (7b type, table VI.3)**

Table VI.5. **Noise barrier costs**

TYPE OF NOISE BARRIER	COST PER SQUARE METRE (US $)
Concrete	75 - 300
Wood	60 -260 (430 for absorptive)
Aluminium or steel (metal)	110 - 240
Metilmet acrylic or polycarbonate (transparent)	250 - 470
Green or vegetative (biowalls)	240 -270
Concrete with New jersey base	125 - 220
Ecotechnic barrier for viaduct	190-215

* for a complete barrier, aesthetic treatment can increase cost from 20 to 40 per cent

As for natural barriers and earthberms, the main cost is often due to land acquisition, while the cost of plants is not high.

VI.6. NATIONAL EXPERIENCE

The following discussion presents data and experiences related to roadway traffic noise barriers as reported by participating countries. It is intended to present examples of several countries' experiences and is not intended to be exhaustive of the experience of all OECD Member countries.

VI.6.1. Australia

Most roadway noise barriers have been constructed out of timber; earthberms have also been used. Other materials that have been considered are fiberglass, aluminum, steel, and concrete (including no fines concrete, Styropor concrete, and glass fibre reinforced concrete). Of these other materials, glass fibre reinforced concrete (GRC) has shown the most potential and has been found to have excellent molding, strength, and durability characteristics and to be lightweight and cost effective. Barriers on bridge abutments or New Jersey crash barriers have been made from GRC.

Figure VI.11. **Typical 2 m timber barrier (reflective 3 type - table VI.3) constructed from CCA treated pine planks 35 mm thick -- Australia**

Noise barriers have to meet the following specifications:

- The barrier must withstand wind loadings for the appropriate terrain.

- Prior to the formal acceptance of any barrier type, the structural elements must be approved by the Road Authority.
- The barrier must have a mass of no less than 10 kilograms per square metre of surface area.
- The overall sound transmission loss through the barrier material must not be less than 30 dB(A) (a test certificate from an approved laboratory is required).
- The barrier must have a design life of twenty years and must be guaranteed for 5 years.
- The barrier must have no gaps or holes in it, or likelihood of them occurring through natural causes, thus allowing noise to pass through.
- The barrier must be designed so that it will not reverberate.
- The barrier must be designed and built in an a manner so that noise will not pass underneath it.
- The barrier must be acceptable from an aesthetic point of view.
- In addition to the preceding requirements, an absorption barrier must have a coefficient of absorption between 0.7 (at 125 Hz) and 0.9 (at 500-1 000 Hz).
- All the components of an absorption barrier must have physical durability with respect to exposure to sun, water, wind, air pollutants and temperature changes.
- The sound absorption materials must have acoustic durability.
- Maintenance requirements of any sound attenuation barriers should be minimal.
- Sound absorption materials should have flame, fuel and smoke ratings that are low enough for them to be used safely beside a highway.

VI.6.2. Austria

More than 500 km of noise walls and earthberms (average height of more than 2 m) have been erected along federal roads. Almost two-thirds of the measures have been walls; the remaining third have been earthberms, combinations of earthberms with walls placed upon them, or steep slopes (supporting structures filled with soil). Walls have normally been built along existing roadway sections (where there is often a lack of space) and on bridges. Earthberms are mainly used when new roadway sections are constructed.

Noise walls have been constructed with the following materials: timber (55%), metal (mostly aluminum) (20%), concrete (molded blocks) (10%), glass (5%), synthetic material or acrylic glass (5%), and New Jersey Profiles (5%).

Artistic design of noise protection barriers meet the purpose in dwelling areas or near towns. In the open countryside, green plantings have proven to be a better alternative for integration into the landscape. Molded concrete frames and troughs, used in combination with trellises, have been used to promote plant growth related to noise barriers along roadways by providing protection from salt spray, ice, and snow lumps.

The proficiency of noise protection barriers (in accordance with German ZTV-LSW 88) is examined according to the following criteria: acoustic requirements; stability and deformation resistance; anti-aging and anti-oxidation properties; stability of dimensions and coloring; resistance to fire and stone throw; maintenance and repair conditions and aspects of traffic safety.

Attempts have been made to apply a special variation of building a wall on top of an earthberm. 1 m-high, earth-filled wooden troughs have been fixed to earthberm crowns without any foundation, saving the cost of a traditional foundation and the purchase of additional property.

Figure VI.12. **Absorptive wooden barrier**
(5c type barrier - Table VI. 3)

Figure VI.13. **Residential area: Artistic design of a concrete barrier (reflective) with ceramic surface elements in Lower Austria**
(8 type barrier - Table VI.3)

Steep earthberms (earthberms with a greater slope as due to an additional artificial supporting structure) have been used where the available space is too small for an earthberm with a natural slope and/or vegetation is desired. Steep earthberms have been found to have higher construction and maintenance costs than normal earthberms and a shabbier appearance in winter months. Concrete, wood, mashed wire, thermoplastics, and used tires have been used for the supporting structures in steep earthberms; guardrails or New Jersey barriers must be provided when steep earthberms are located close to the roadway. Experience from the past 10 years has shown that steep earthberms are extremely difficult and costly to maintain and that they should, therefore, be used sparingly.

New Jersey barriers have been used along elevated roadways to provide noise abatement. They can be erected separately, in combination with other barriers, or as a staggered barrier. A reduction of 3-4 dB(A) can be achieved with a modified 1 m-high (not the standard 80 cm height) New Jersey barrier. Staggered barriers have used when a height greater than 1.5 m has been required. This has normally done in one of two ways: the bottom part of the noise barrier behind the New Jersey profile was left open, thus permitting plants to grow and fill the gap between the New Jersey barrier and the noise barrier, or the space behind the New Jersey barrier was filled with earth, and vegetation was established to a height 80 cm to 1.0 m above the roadway. Staggered barriers cost more than standard noise barriers.

Figure VI.14. **Reflective barrier with landscape integration by the use of (1) transparent elements and (2) landscaping (Austria) (5f type - Table VI.3)**

VI.6.3. Denmark

Great emphasis has been placed on the visual quality and the setting of noise barriers for both motorists and residents. The visual effect of high noise barriers has been improved by using colors (i.e., barriers have been painted in horizontal color lines, with the light colors at the top and the dark colors at the bottom), by using transparent materials, by inclining or sloping the barriers away from the roadway, or by planting low vegetation in front of the barrier and higher vegetation behind the barrier.

Figure VI.15. **3 m high noise barrier (reflective) made of glass in Soborg Mainstreet in Gladsaxe, Denmark. The glass has the effect of making the barrier look very light (8 type and 5f type barrier)**

If the desired height of the barrier has been more than 4-5 metres than earthberms have most often been easier to fit into the landscape than noise barriers. The earthberms have primarily been used in natural surroundings and in less densely populated areas, as the earthberms require relatively large amounts of space.

VI.6.4. Finland

The Road Administration has annually built approximately 10-15 km of noise barriers. Most barriers have been made from wood. In 1989 the Environment Ministry published a report on the efficiency of 8 noise barriers in the Helsinki area.

Measurements were made by registering simultaneously 5 minute levels at a reference point 0.7-1.0 m vertically above the barrier, and at target points, at the levels 2 and 4 m above ground, behind the barrier. 1 to 3 target points were used, at different distances from the barrier. Short barriers were

Figure VI.16. **Low noise barrier (absorptive) on motorway centerlane (length 440 m, height 1 m) -- Finland**

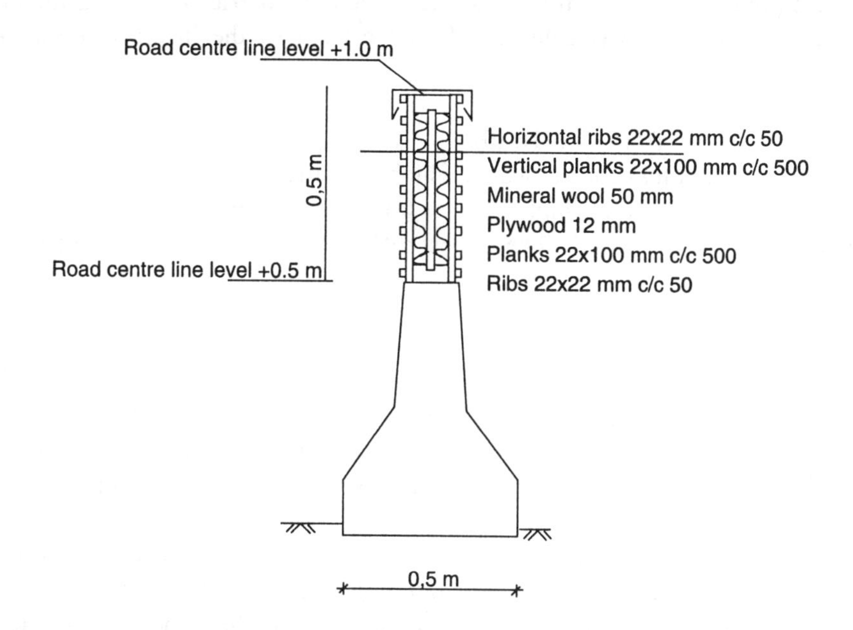

a 0.5 m wooden barrier (steel uprights, on both sides horizontal wooden ribs on vertical planking, rockwool acoustic absorbent and in the middle plywood) raised on 0.5 m concrete New Jersey-type crash barrier elements (7b type of table VI.3)

Table VI.6. **Comparison of different types of barriers (Finland)**

Barrier Type	Barrier height (m)	Targeted Point distance (m)	Insertion loss (db) *	Barrier attenuation	Calculated effect **
Wooden elements on bank	3+1	13 37	13/10 10/10	13/10 12/11	16/10 15/12
Concrete elements	2	16 40 70	11 10/10 7/8	11 14/11 12/10	10 11/8 10/8
Wooden elements	2.7	14 25	11 11	11 12	11 12
Wood on concrete	3	15 27	11/9 10/9	11/9 11/9	13/9 13/10
Wooden elements	2.7	20 33	10/8 10/10	11/8 14/11	14/9 11/10
Wooden elements	2.7	17 50	12/10 8/8	13/10 11/9	13/10 8/8
Wooden elements on bank	2+1	18 50	14/11 11/10	15/11 14/12	13/10 11/9
Earth embankment	5	25 43	16/14 16/13	17/14 18/13	20/18 18/16

* Target point height 2 m / 4 m, if only one number, 2 m.
** Calculated in accordance with the Nordic model.

measured at their midpoint. Insertion loss was determined as the difference between the noise level measured at the reference point and the level measured at the target points, corrected by factors dependent on target point distance from the barrier, reflections, barrier form, and the effect of ground attenuation. Barrier attenuation was determined on the basis of the insertion loss. The results are presented in table VI.6.

The insertion losses vary from 7 to 16 dB(A). Target points near the barrier (distance 13-18 m) gave 11 to 15 dB(A) insertion loss for the walls, which is a small variation. Calculation predicts lower losses for < 3 m barriers and generally higher losses for barriers over 3 m than the measurement results show, but differences are usually small. The largest difference is 4 dB(A) (for the earth embankment).

VI.6.5. Italy

Many materials have been used for noise barriers, including steel or aluminum, wood, concrete, polycarbonate, and refractory or ceramic materials. Barriers are generally divided into types according to the materials of which they are made, along the lines of the scheme set out in table VI.3. There are:

- *"Natural" (vegetation) barriers*, made up of vegetation that is special both in plant type and in layout on the infrastructure or the surrounding areas. This can include the earthworks (anti-noise embankments) that form their base in those cases where they are not an integral part of the infrastructure.

- *Artificial barriers*, composed of panels of various materials, such as concrete, wood, aluminum, steel and transparent plastic materials (e.g., polycarbonate and methacrylate), mounted on supporting frames set in the ground or connected to the engineering structures.

- *Mixed barriers*, made up of artificial supports (concrete, steel, wood) that allow the growth of vegetation. Several types of mixed barriers (an artificial load bearing structure covered with vegetation) are commercially available and can be divided into three categories: (1) biowalls, made up of load bearing elements in concrete, wood or other materials, containing and enclosing earth, in which vegetation is planted; (2) vegetation barriers, composed of artificial absorptive panels, completely covered with vegetation; and (3) green walls, composed of an artificial metal cage structure, filled with a mixture of earth, in which climbing vegetation is planted.

Great importance is attached to the choice of the most suitable species of vegetation to be used in the various climatic zones. The species are chosen not only for their phytotechnical characteristics, i.e. the capacity to resist erosion or mechanical actions in terrains with potential landslides and the capacity to consolidate the earth through their roots, but also for their resistance to the environment -- drought, rain, mist, etc. -- as well as their ability to absorb such polluting agents as: fumes, particulates, heavy metals. This assures that the anti-pollution action of the noise barrier is being enhanced.

The following list of plants may be of interest:

Northern Italy (rainy climate)	**Southern Italy (droughty climate)**
• Acer/pseudoplatanus	• Cupressociparis leylandi
• Tilia platyphyllos	• Chamaechyparis lawsoniana
• Carpinus betulus	• Quercus

- Fagus sylvatica
- Quercus robur
- Pinus pinaster
- Eucalyptus sp.
- Populus nigra piramidalis

<u>Central Italy</u> (mixed climate)
- Acer pseudoplatanus
- Chamaeciparis lawsoniana
- Cupressociparis leylandi
- Populus berelinensis
- Tilia platyphyllos
- Quercus ilex
- Carpinus betulus
- Fagus sylvatica
- Quercus robur
- Pinus alepensis
- Pinus pinaster

Some general rules are noteworthy:

♦ It is always preferable (if possible) to integrate the natural barrier with other elements, possibly "natural" as well.

♦ Tree species must never be used alone, but must always be integrated with bush or shrub species capable of filling the empty spaces at the base of the trees, or the gaps between the trees, as in the case where cotoneaster bushes fill the gaps between Austrian pines.

Figure VI.17. **Biowall on cutting (absorptive) Biowalls are made up of load bearing elements in concrete, wood or other materials, containing and enclosing earth, in which vegetation is planted - Italy (4 type - table VI.3)**

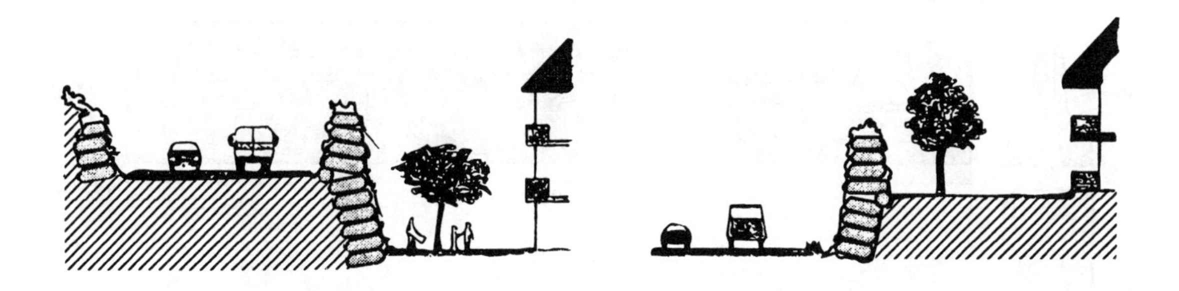

♦ The depth of the mixed vegetation belt should (preferably) not be of less than 6-7 metres, with a layout that begins as close as possible to the noise source (edge of the infrastructure). This "closeness" is meant in relative terms, since at least 3-4 meters must be left as a fire-guard.

- For a further increase of protection levels it is possible to insert artificial panels, which better the uniformity in time of noise abatement, and function alone in the first growth stage of the plants.

Figure VI.18. **Prefabricated Green Wall (absorptive). Green walls are composed of an artificial metal cage structure, filled with a mixture of earth, in which climbing vegetation is planted - Italy (4 type - table VI.3)**

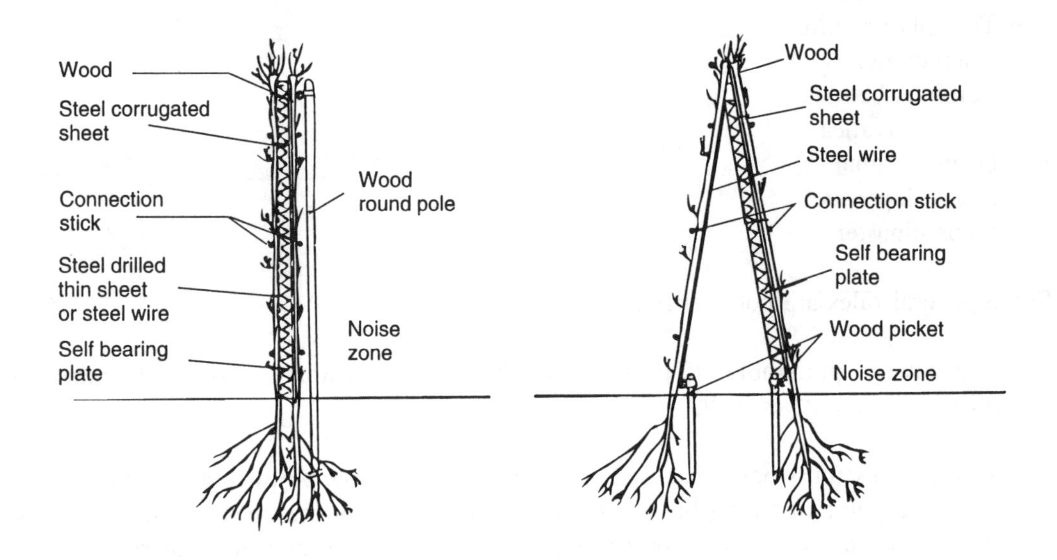

Figure VI.19. **Sound-absorbing New Jersey Barriers - Construction details of the phonoabsorbent insert (7c type - table VI.3) - Italy**

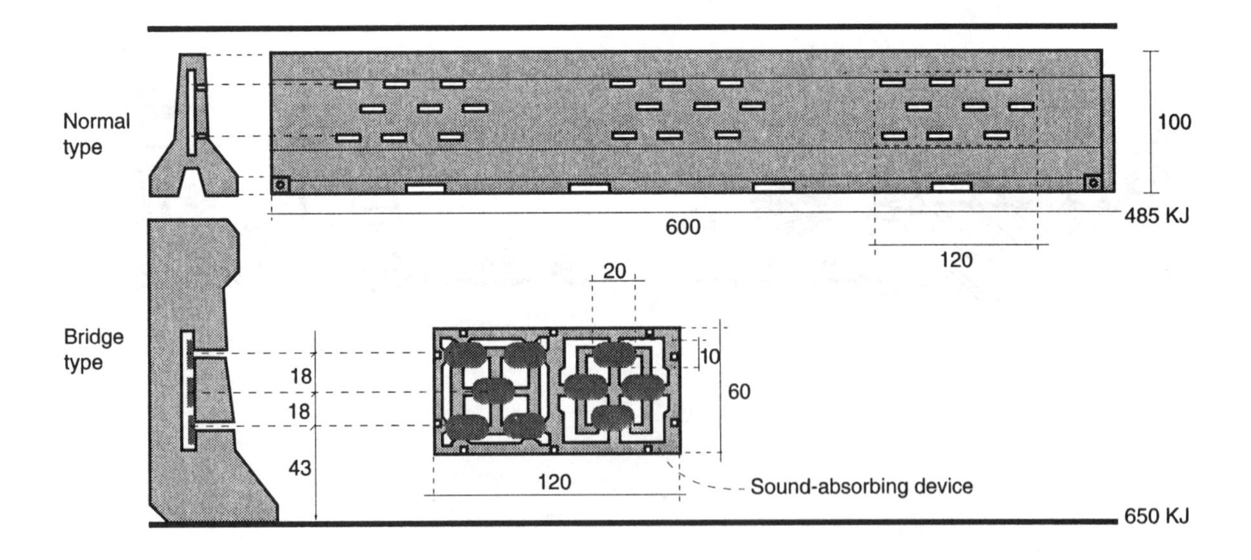

Artificial barriers in Italy have been principally studied for existing roads, for which criteria of high efficiency and small utilisation are paramount. This explains the interest in barriers with New Jersey sound-absorbing basements which assure:

- A single structure for traffic protection and a sound barrier with a New Jersey type basement (type 7 barrier in table VI.3);

- Specific sound-absorption for low frequencies (up until 250 Hz) to complement the absorption characteristics supplied by the panels mounted on such basements.

As noted in Chapter IV bridge protections from vehicle impacts are a delicate problem. The bridge type barrier of figure VI.19 is crash tested for 650 kJ in comparison to the normal barrier (485 kJ). The sound absorption of low frequency uses the damping of incident sound waves through the use of resonators placed inside the barrier for the purpose; a series of ducts of adequate size connect the resonators with the outside.

VI.6.6. Japan

Over 200 km of noise barriers have been built annually, and a total of 3,100 km of barriers will have been constructed by the end of Fiscal Year 1992. The first barrier was made from concrete. Concern over reflected noise from reflective concrete barriers led to the development of sound absorbing barrier panels. Currently, the most popular type of panel consists of glass wool wrapped in plastic film and covered with aluminum plates.

Figure VI.20. **Noise reduction effect by noise reducer**

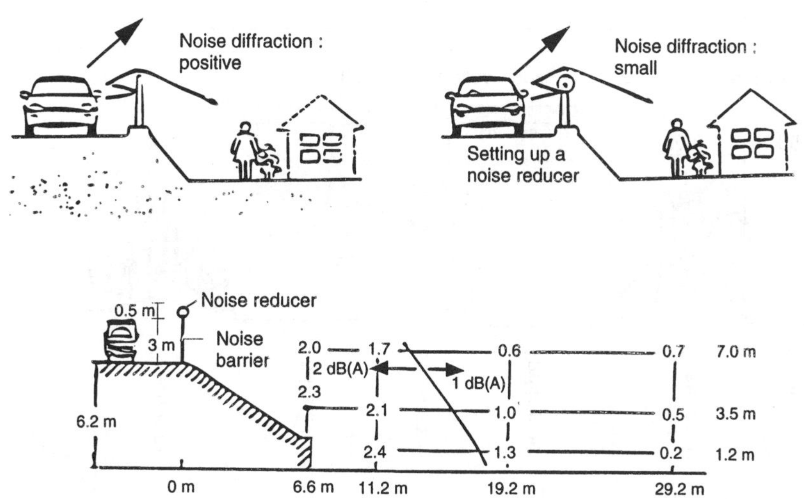

Remark : Setting up a noise reducer improves the effect of noise reduction near a noise barrier

One recent innovation for noise reduction has been the "noise reducer." This is a sound absorbing body set up on the top of a noise barrier to reduce noise without increasing barrier height. "Noise reducers" are currently being tested on various expressways.

Figure VI.21. **Example of setting up a noise reducer**

Figure VI.22. **Standard cross section of Tokyo outer ring road**

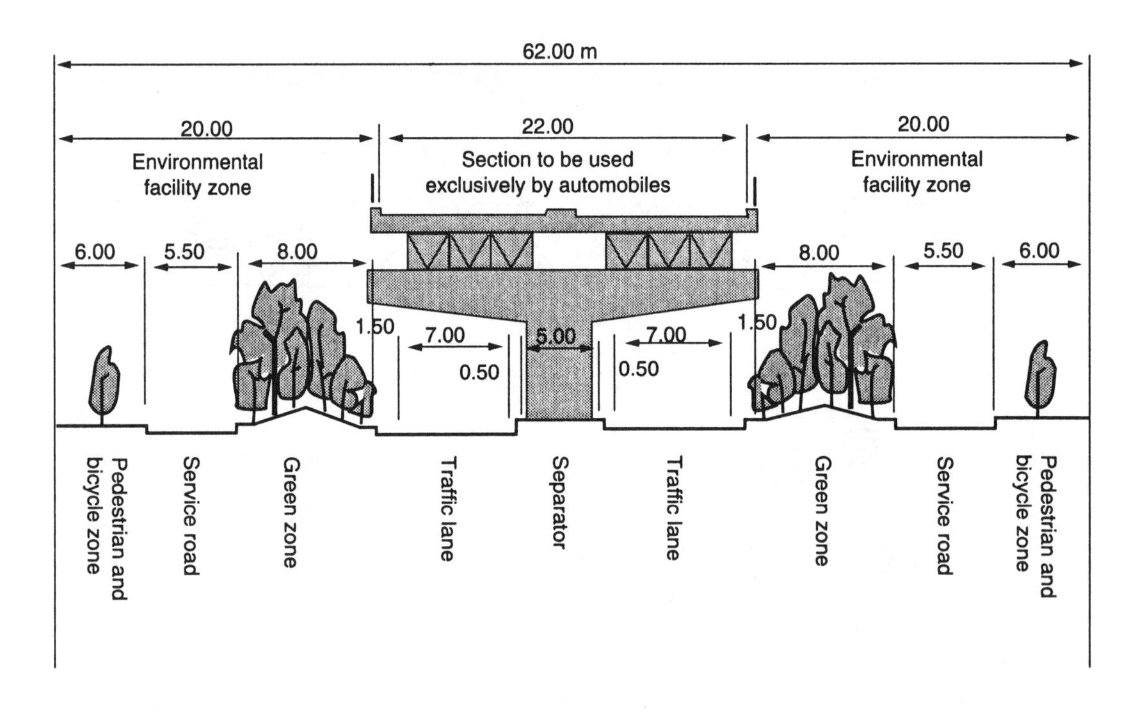

VI.6.7. Netherlands

Barriers tilted away from the roadway have been used to prevent noise reflection to the other side of roadway. Investigations are being made to evaluate the improvement in effectiveness and practicality

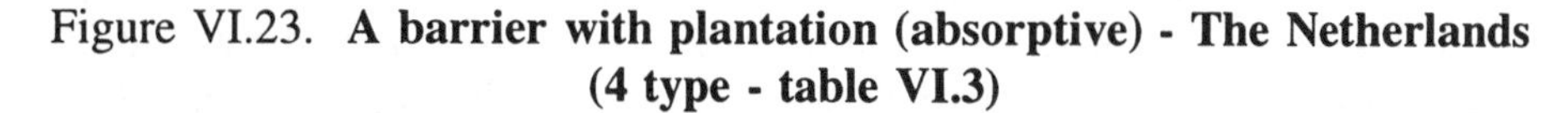

Figure VI.23. **A barrier with plantation (absorptive) - The Netherlands (4 type - table VI.3)**

Figure VI.24. **High-absorbing concrete barrier (absorptive) - The Netherlands (5b type - table VI.3)**

achieved by slanting the barrier towards the roadway and using porous asphalt pavement. Noise barrier height has been limited to 5 metres, producing a maximum noise reduction of approximately 10 dB(A). If a greater noise reduction is desired, a partial canopy can be constructed over the roadway. One such canopy, with a length of 1.600 metres, has been constructed.

Gaps have sometimes been left at the bottom of noise barriers so that (1) no special drainage facilities are needed, (2) smaller animals have unrestricted freedom of movement, and (3) maintenance costs are lowered since there is no contact between the soil and the noise barrier. An acoustic study of the effect of gaps has shown that an opening 20 centimetres wide reduces the noise reduction of the barrier by less than 1-2 dB(A).

VI.6.8. Norway

Originating from Norwegian tradition to erect fences around the houses, there have been three types of roadway noise barriers:

1. An **area barrier** screens an area that is not part of the roadway's function. The barrier is part of the area it screens and not part of the road.

2. A **site screen** shields outdoor areas (gardens, parks, yards) where the area and roadway/street are part of each other's function. Houses/buildings are usually close to the roadway. Although the screen is part of the roadway/street area, it is designed to be part of the housing/building area. The screen can be erected as an extension of a row of houses.

3. A **local screen** only shields part of an outdoor area and will not provide noise abatement for the building itself. The screen can be connected directly to the house or be detached.

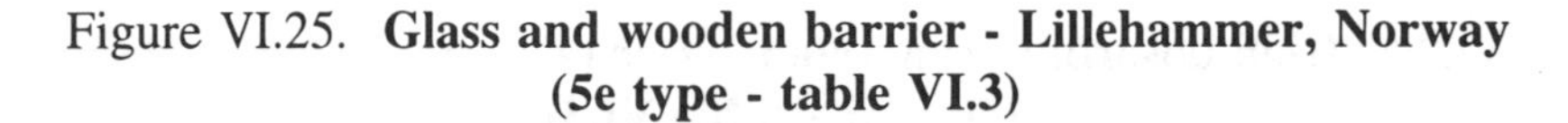

Figure VI.25. **Glass and wooden barrier - Lillehammer, Norway (5e type - table VI.3)**

Norwegian highway standards and specifications classify roadways in various categories depending on the function of the roadway and the character of its surroundings - main roads, collector roads, and access roads in densely, average and scattered developed areas. A noise barrier's location and height depends on both its acoustic effect and its aesthetic effect. The following heights and lengths of various types of barriers have been suggested in table VI.7.

Table VI.7. **Suggested heights and lengths of various types of screens - Norway**

TYPE OF SCREEN	1	2		3
		SITE SCREEN		
HEIGHT LENGTH REQUIREMENT	AREA BARRIER	a)GARDEN SCREEN	b) CITY SCREEN	LOCAL SCREEN
Max. Length with variation **Max. Length without variation**	1 km 100 m	500 m 50 m or max. 3 properties	From house/ building to house/ building	5-10 m (extension or detached)
Screen height (m)	2-4	1.5-1.8 (2.1)	2-4 (5) adapted to building	1.5-2.5
Embankment height (m)	4-5 slight slope otherwise ≤ 3	1-1.5 or fill-in towards houses/ buildings	Not relevant	Not relevant
Distance of screen to houses/buildings	≥ 10 m	≥ 7 m	0 m	Connected to houses/ buildings
Minimum distance of screen to edge of road (see also items c, d, and e)	10-15 m for screen length ≥ 500 m otherwise 7-10 m	On the curb	Screen level with houses/ buildings	(≥ 3 m)

Almost every type of barrier has been built. The first barriers, built of wood, cracked due to poor installation and required much maintenance. Concrete, has become the dominant barrier material. The concrete barriers have been better in terms of maintenance, but there have been problems due to poor foundation work, poor sealing of element joints, and poor landscaping. Double-sided wooden barriers, containing absorbing mats, have also been used. Green and "living" barriers have been tried but are still in the test stage, due to cold winters and the danger of dehydration.

VI.6.9. Spain

Spain has used concrete noise barriers with areas to hold soil, making it possible to use plants to hide the barrier faces and offer the sensation of a natural barrier.

The selection of the materials is conditioned by two acoustic properties required of screens: their sound transmission performance and the absorption qualities of the screen as a whole.

Figure VI.26. **Artificial barrier - Spain**
(5d type - table VI.3)

Standard ISO R-144 is used to define the acoustic isolation index R, and screens are required to have an R transmission abatement index equal to or in excess of 25 dB (A) for the standardized road noise spectrum. Until a European standard is issued, the German standard ZTV-LSW 88 is used to classify the absorbent characteristics. Fibre mineral wool or perforated panels, with appropriate orifice sizes, are used as the main absorbent materials.

To avoid the effects that can be caused by multiple reflections when screen are installed on either side of a road, the recommendations made by the "Centre Scientifique et Technique du Bâtiment" (C.S.T.B.) of Grenoble are followed:

H > L/5: Absorbent material should be used in the screens.

L/5 > H > L/10: The decision to use absorbent material depends on the environment and on the possibility of sloping the screens. The efficiency of the two solutions should be studied.

L/10 > H > L/20: Sloping screens are to be preferred to the use of absorbent materials, because generally they are more effective.

H < L/20: The use of absorbent material or sloping screens hardly affects the final result.

H = Height of the screens
L = Distance between two screens located opposite one another

Figure VI.27. **Artificial barrier "total screen" - Spain (9 type - table VI.3)**

VI.6.10. United States

As of the end of 1989, over 1,160 kilometers of noise barriers have been constructed. The following table shows total noise barrier lengths by material type.

Table VI.8. **Total noise barrier length by material type**

Single Material Barriers		**Combination Barriers**	
Material	**Length in kilometres**	**Materials**	**Length in Material Kilometres**
Block	370	Berm/Wood	35
Concrete/Precast	238	Berm/Concrete	31
Berm only	81	Wood/Concrete	27
Wood/Unspecified	63	Concrete/Brick	20
Wood/Post & Plank	59	Wood/Metal	12
Concrete/Unspecified	48	Metal/Concrete	11
Metal/Unspecified	44	Berm/Block	11
Wood/Glue	40	Concrete/Block	10
Laminated	11	Wood/Block	7
Brick	12	Berm/Metal	6
Other		Berm/Wood/Block	5
		Berm/Wood/Metal	5
		Other	17
Total	**966**	**Total**	**197**

Almost three-fourths of the barriers that have been constructed range in height from 3-5 metres. The overall average barrier height is 4 metres; only 3 per cent are more than 6 metres tall. The only barriers over 9 metres in height are made of concrete. Barriers have been constructed from materials that include concrete, masonry block, wood, metal, earthberms, brick, and combinations of all these materials. The apparent popularity of block shown in the above table is due to the fact that the State of California almost exclusively uses block material and has built over one-third of all noise barriers.

Figure VI.28. **Precast concrete panels with graphic designs adjacent to a playground (reflective) - United States (8type - table VI.3)**

Average unit costs for all barrier materials range between $110-180 per square metre, except for earth berms which average only $30 per square metre. Average unit costs for the years 1986-1989 for all barrier materials have remained fairly constant, ranging from $100-280 per square metre (earth berms $50-80 per square metre). Overall average unit costs for masonry block and concrete, materials which comprise over 56 per cent barriers built, are $170 per square meter and $140 per square meter respectively. Metal costs are comparable to those of wood ($110 per square meter).

The most notable trend in highway traffic noise barrier construction has been a dramatic increase in the amount of construction in the two most recent years of reporting, 1988 and 1989. Expenditures for these two years have almost tripled the average yearly expenditure for 15 years of recordkeeping. Most barriers have been made from masonry block or concrete, range from 3-5 meters in height, and average $140-170 per square meter in cost.

The majority of noise barriers built in the United States have hard, sound-reflecting surfaces. However, sound-absorbing barriers have been used where reflections from single or parallel noise barriers were of concern. Tilted noise barriers (tilted 5-10 degrees from vertical) have been used as an alternative to sound-absorbing barriers to prevent reflections by deflecting the sound waves upward. Transparent noise barriers, constructed using polycarbonate panels, have also been used in situations where an unimpeded viewscape has been desired. Periodic cleaning of the panels has been cited as a maintenance requirement.

VI.7. CONCLUSIONS

Noise barriers can be constructed as earthberms, walls, or walls on earthberms. A noise barrier can reduce roadway traffic noise levels by 10-15 dB(A) in the area immediately behind the barrier. Vegetation provides especially a psychological effect, reducing annoyance to traffic noise but providing only limited reductions in roadway traffic noise levels.

Noise barrier design should include appropriate consideration of:

- the noise level reduction,
- aesthetics,
- safety,
- maintenance,
- drainage, and
- the cost of construction and maintenance.

The cost of construction for noise barriers varies widely among reporting countries, i.e., $60-470 per square meter (US $) for materials that include concrete, wood, metal, acrylic or polycarbonate, green or vegetative, and New Jersey.

Aesthetic design and the integration of noise barriers into the landscape and the environment are of special importance. This is especially true of barrier height, the choice of material, and the barrier shape, structure, and colouring. A successful design approach for noise barriers should be multidisciplinary and should include architects/planners, landscape architects, roadway engineers, acoustical engineers, and structural engineers.

Community participation in decisions regarding noise barrier provision and design must be systematically assured.

Earthberms:

- are typically covered with vegetation,
- have a very natural appearance,
- are usually attractive,
- typically allow more sunshine (less shade) and better air circulation than walls,
- can be used as a means of disposing of excess earthen material,
- normally do not require safety protection (guardrails) for vehicles that leave the roadway,
- are usually less costly to install and maintain than walls, and
- offer a practically unlimited useful life.

However, construction of earthberms can require extensive areas of land and considerable utility relocations.

Walls take less space than earthberms and can be constructed with wood, concrete, masonry, metal, and other materials. Reflective walls can provide benefits for residents behind the walls, but can increase noise along the roadway for motorists. Several countries report noise increases for residents that live directly opposite walls, while other countries report only a perceived (not an actual) noise increase for these residents. The effectiveness of parallel reflective walls (a wall on both sides of the roadway) can be reduced if the walls are tall and close together, i.e., a narrow roadway cross-section. Reflective walls can be inclined or sloped to direct noise upward into the atmosphere and reduce or negate reflected noise problems. Walls can also be constructed with absorptive surfaces to reduce noise reflections.

CHAPTER VII

INTEGRATION OF MEASURES AND COSTS

VII.1. THE INTEGRATION OF PROTECTION SYSTEMS

The study of the various possibilities of acting on road noise emission and diffusion characteristics has, at this point, clearly showed that it is always possible to improve the acoustical performance of the infrastructure. The real problem is evaluating the combined effect of the various possible solutions, since solutions do not necessarily add together, even when they seem complementary. It is also important to evaluate the true cost of abatement of traffic-induced sound energy.

This chapter attempts to address these questions or at least provide useful indications, and points out that the economic factors needed for evaluation must be obtained from a scientific assessment, driven by many elements.

The focus here is not on infrastructure in tunnels or cuttings, having already seen in Chapter IV that these are, by their nature, "euphonic". The interest is on "open air" infrastructure, whether raised or on viaducts. On roads outside urban areas -- or in any case on roads without nearby buildings -- it is possible to focus on:

- safety barriers,
- pavements,
- noise screens,
- covering plants.

When operating instead in an urban area, not classifiable as a "free" road (such as a "U" shaped road, surrounded by buildings), possible action is even further limited to:

- pavements;
- insulation of facades;
- screens and barriers (if this is compatible with the use of houses).

Figure VII.1 illustrates a graphic case of combined protection systems on a raised, open road. Obviously, the various elements contributing to noise abatement cannot be simply added together (see Chapter I).

Figure VII.1. **Euphonic road**

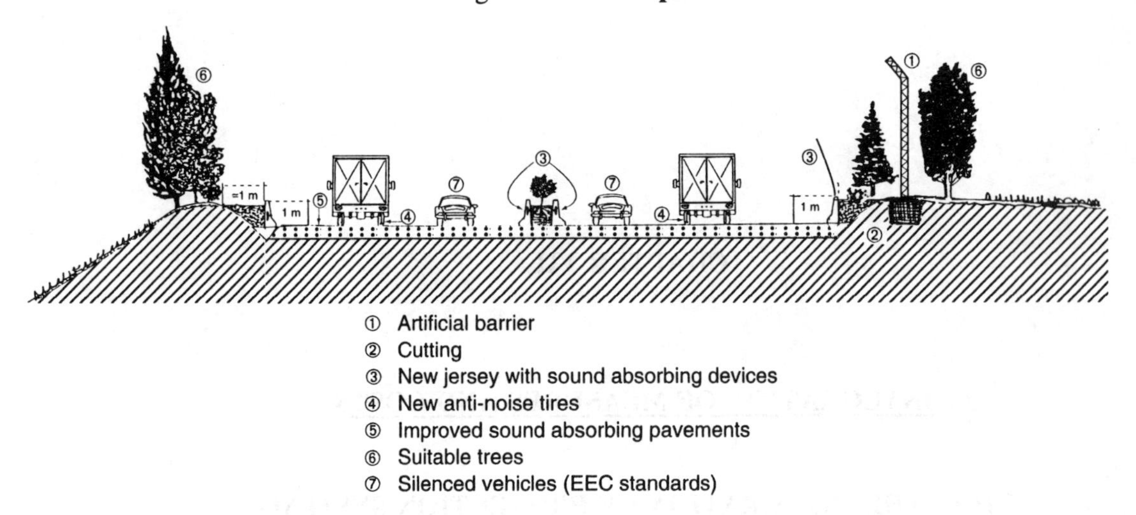

Best results can be obtained with the calculated combination of the different anti-noise devices. Mathematical models must be used in advance for this evaluation. One problem for the estimate arises from the change by the sound-absorbing pavement to the tyre/road noise frequency spectrum. The other absorbent devices must be calibrated with the changed spectrum.

It is possible to act on pavements at different levels of sound control (see chapter V) affecting different frequency bands. Lateral and central safety barriers further reduce noise (in frequency bands that can be different from or equal to those of the pavement), through their sound insulation and absorption characteristics, resulting from the cavities within them.

In addition to this is the effect of the vegetation on the small side hills, augmented by that of the vegetation covering the lateral slopes of the raised surface; all of which can be further augmented by screens placed within the vegetation or on top of the safety barriers.

It must be noted, however, that the noise abatement obtained is certainly not the sum of the reductions provided by the individual measures, and, among other things, the cost of each reduction is different. Only a careful combination of effects, taking into account the frequency spectra of the reduced noise, will allow precise evaluation of the effectiveness of the actions taken.

Above all it must be remembered that not all of the reductions indicated in the figure are "certain" and lasting without proper action:

- Vegetation is effective only if there are no interruptions in the protection elements, and if the plant "facies" are constant throughout the seasons;

- Pavements must be laid with constant quality and, with the passing of time, must be properly maintained.

It must also be mentioned that the combination of various effects must be properly evaluated. One of the discussions, most current in all OECD countries, is the combination of artificial screens and draining pavements.

VII.2. COMPLEMENTARITY OF NOISE BARRIERS AND LOW-NOISE PAVEMENTS

Today sub-urban areas are most relevant for the use of noise barriers. Here, it is possible to benefit from the excellent complementarity of screens and low-noise pavements. The optimisation of the combination of noise barriers and low-noise wearing courses can allow a reduction of barrier heights in some cases, thereby reducing their cost, or alternatively be made to yield greater noise reduction. Naturally, for a "U" road, the complementarity disappears, and it is necessary to use:

- Either euphonic pavements (which eliminate all traffic noise frequencies)

- Or structures (where possible) such as artificial tunnels, or barriers with upper sound-absorbent elements suspended over the road (such as the baffles shown in Chapter IV).

The solutions described are always complementary, either through the frequency band (in the case of the euphonic pavement) or by the physical area protected (the top floors in the case of baffles). When combining these protection systems the energy content of each spectrum band should be taken into account. This makes it possible to match the characteristics of the barriers and/or the baffles to those of the particular pavement used.

If the pavement surface reduces sound levels in the medium and acute ranges, it is necessary to design a barrier that reaches optimum efficiency in the lower frequency ranges. It is therefore a problem of global acoustical engineering -- and of civil engineering -- to produce the required barrier which must also be stable, durable, and inexpensive. Throughout this report a large number of solutions to this type of problem are put forward, and more can be found, once an evaluation approach has been correctly established.

"Ecotechnical" safety barriers, New Jersey barriers with cavities of several decimeters, and concrete block walls are all examples of barriers that are complementary in the frequency levels affected, using draining pavements.

VII.3. ECONOMIC ASPECTS

The economic approach the Group has attempted to set out in this chapter is not absolutely deterministic:

- Because costs evolve irregularly in time everywhere and because, alongside the average figures tied to the common conditions of production and installation of the products, there may be considerable differences arising from local factors specific to each of the different countries;

- Because the techniques in question are highly innovative (sound-absorbent pavements) or have markets with small volumes (road noise barriers), they are also strongly affected by the conditions of their application. Consider for example the great variety of foundation types available for barriers that have identical materials and acoustical properties. The cost variable therefore depends on the context and the situation where the noise barrier is installed. The costs indicated below are therefore to be taken as indications only, to help focus ideas on the topic.

In spite of a certain inevitable inaccuracies, the Group has provided some elements to:

- identify the range of variability of sound-absorbent pavements;
- compare costs of other techniques for reducing road noise.

VII.3.1. Costs of low-noise pavements

Chapter V reviews the multitude of road pavements having anti-noise qualities. It is however widely believed that only a few of these can be considered truly specialised for use in noise control (macroporous types B, C, and D of table V.1). The other (microtexture) pavements can only be classified as low-emission pavements and as such do not have a significant impact on the global reduction of traffic noise, also because they operate in the high-frequency ranges, similar to those of the more common sound-absorbent barriers.

Obviously there is no general agreement on the above statements. It is however always possible to check the actual effect of different solutions in specific applications, setting aside the above assessment, which must naturally remain highly generic.

Costs of Microtexture Pavements

In this sector there is a wide range of variables, due to the different use of thin wearing courses with hot-laid bituminous mixes (with or without modified bitumen) and the more expensive surface treatments, often used in urban areas, especially where accident rates are high. The following table sets out a series of costs in current US Dollars, indicating the maximum and minimum levels obtained from available information (European OECD countries)

Table VII.1. **Cost of microtexture pavements**

Type	**Thickness (cm)**	**dB(A) reduced**	**Cost $/m²$ (min-max)**
Fine bituminous mix with natural binder	2 - 3	1	2 - 2.5
Thin layer of bituminous mix with modified binder	1.5 - 2	1	3 - 5
Surface treatment with calcined bauxite and Shell Grip-Spray epoxy binder (U.K.)	0.5 - 1	1 - 2	12.5 - 19
Surface treatment with artificial chipping and epoxy binder (Griproad - D, I)	0.5 - 1.2	1 - 2	8 - 10.5
Concrete slab with exposed surface aggregates	3 - 4	3 - 5*	2 - 3

* compared with the traditional cement concrete pavement

As can be seen from the table, the results obtained are negligible, although costs are moderate, except for high-grip wearing courses, which are however generally used mainly for safety reasons.

Costs of Bituminous Draining Mixes ("Drainage Asphalt) Used as Wearing Course

The major interest is on this mix and a few other types, since its wide use and dissemination makes it possible to carry out an economic evaluation based on the results obtained. The other types of "microporous" pavements mentioned are still in the experimental stages, and so the costs described later are not yet final (they should, most likely, decline).

The price of bituminous draining mixes has been declining for some time. The focus is on the more widespread, high-performance mixes, made of low-abrasion aggregates, modified bitumen with high quality elastomers, 4 cm thick and 20 per cent void content, laid on a tack coat, also of modified bitumen. Reference is made here to French data, but the figures are comparable to those of other European countries with higher levels. In 1986 some projects had costs of $ 10/m2, but over time the cost levels have declined (see table VII.2).

Table VII.2. **Evolution of the cost of bituminous draining mixes**

Year	1987	1988	1989	1990
$/m²	5.5 - 8	5.5	4 - 4.8	4

Currently costs are stable (figures refer to finished works), aside from a few variations tied to local logistic problems and the size of the job. Sometimes extra costs are incurred through the difficulty of procuring aggregates or the use of higher-grade elastomers; in these cases costs can reach $ 7-8 per square metre. Table VII.3 sets out a statistical distribution of the work site costs encountered by the French highway contractors; in 1992 prices ranged between 3 and $ 12 per square metre, taking into consideration all types of bituminous draining mixes (including those with ordinary binders) and all thicknesses.

Table VII.3. **Distribution of work site costs in France**

Cost per square metre ($)	< 4 $	4 - 6 $	6 - 8 $	> 8 $ (1)
Percentage of costs recorded	13%	60%	17%	10%

(1) applies to special work sites; steep slope sites and experimental sites

As a reference cost for bituminous draining mixes we will therefore take 6 $/m^2.

The costs of thick porous pavements have not yet been defined: in some French applications (such as the CETE of Lyon), costs have reached $ 70/square metre, but this case involved complete reconstruction of the pavement, and not just of the top layer, providing extremely good absorption effects even between 200 and 500 Hz (with thicknesses greater than 10 cm).

Working with layer thicknesses of 10 cm, other experiments and applications have established a cost of approximately $ 16 per square metre. The presumed cost per square metre of a D type euphonic pavement is approximately $ 70.

VII.3.2. Elements of comparison with the costs of other specialised anti-noise protection systems

The noise production (and containment) chain is shown in figure VII.2, which schematizes the various parts involved and shows that it is possible to intervene on all parts of an acoustical "chain", from the source (vehicle) to the receiver (the inhabitant of a structure located along the road).

Figure VII.2. **Acoustical chain**

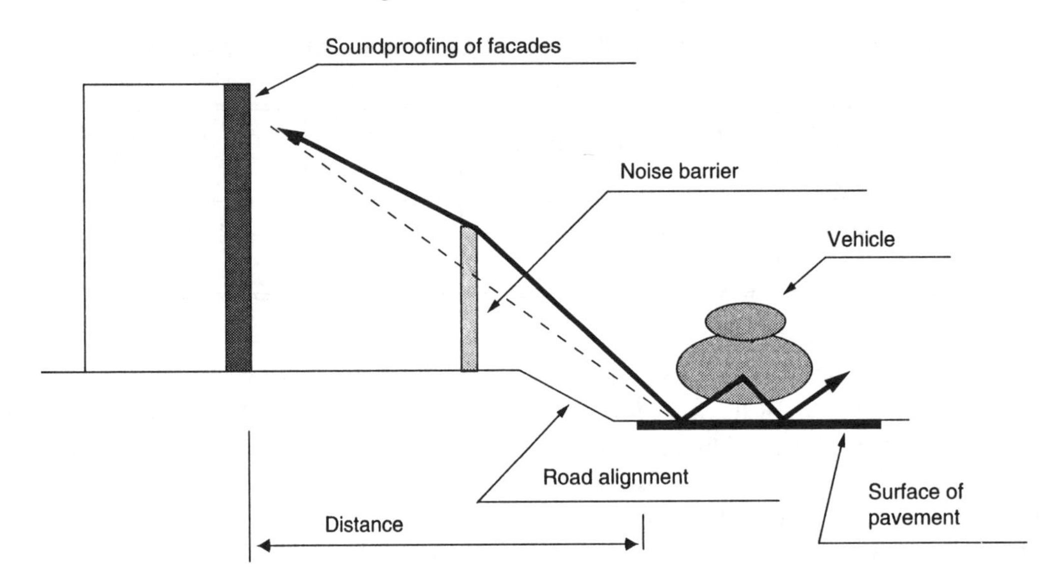

Cost of Pavements

We have examined the costs of pavements that provide reductions in the range of 1 to 2 dB (A) (for microporous pavements) to verified 3 dB(A) for draining wearing courses 4 - 5 cm thick, with the possibility of reaching 5-7 dB(A) reductions by altering thickness and voids content. Obviously it is necessary to add the management costs to the installation costs when they are different from those encountered with normal pavements. This is especially true with regard to the problem of clogging in bituminous draining mixes, which in urban contexts requires unclogging at least every two years. Some countries avoid the use of bituminous draining mixes for this reason, since the special machines used for this operation do not always provide satisfactory results. The annual cost is approximately $ 1.3 per square metre, to be added, converted to net present terms, to the installation costs. After eight years the bituminous draining mixes can be recycled and regenerated on site at a cost of $ 5 per square metre, after which it is necessary to begin washing again. The present cost of these operations, distributed over a period of 16 years, is therefore $ 14.85 per square metre.

In any case, even a noise reduction of 5-7 dB(A) does not offer sufficient protection in some cases, since house facades are subjected to noise levels of 72 to 83 dB(A) and it is necessary to reduce the level to 65-60 dB(A) (and even less at night). Noise levels must be reduced by 12-18 dB(A) in common cases and 22-23 dB(A) in difficult cases.

Figure VII.3. **Protection of top floors : "Total screen"**
Double artificial barriers with baffles (absorptive elements) put over the road (see table VI.3)

Cost and Effectiveness of Other Protection Systems

We do not consider it relevant to discuss the option of changing road alignment or layout -- i.e. change from a viaduct to a tunnel or from embankments to cuttings -- for acoustical reasons only. This could cost several million $ per km, therefore becoming a political, not an economic choice. The only specific techniques to be evaluated from an economic point of view are the construction of noise barriers and the insulation of facades.

A sound-absorbent or insulating barrier costs between 600 and $ 1 600/m of road to protect one side of the roadway: the variability is dictated by the height (3 or 4 m) and the more or less costly foundation. If protection is required on both sides of the road the cost doubles, to $ 1 200-3 200 per metre.

The use of taller noise barriers increases costs, which must be evaluated on a case-by-case basis. A useful parameter can be the example of a case of total protection (efficiency 15-25 dB (A)) obtained by a double wall of screens, with baffles (sound-absorbent elements) suspended over the centre of the roadway to protect the upper floors of the buildings. For this case the cost is $ 4500 per metre.

Table VII.4. **Efficiency and cost of acoustical protections**
(estimated useful life for all devices: 15 years)

Type of acoustical protection	Average efficiency[1]	Cost per linear metre of road	
		Protection of one side of the road $/m	Protection of both sides of the road $/m
Noise barrier (with "normal foundations")	6 - 12 dB(A)	600 - 1600	1200 - 3200
"Total" screen *	15 - 25 dB(A)	-	6600
Noise absorbent surfacing with drainage asphalt (°) (type B)	3 - 5 dB(A)	-	120 (297) ***
Optimised pavement (porous, semi-thick) (type C)	5 - 7 dB(A)	-	less than 1400
Anti-noise (euphonic) pavement (type D)**	5 - 7 dB(A)	-	1400
Improving soundproofing of facades ****	5 - 10 dB(A) *****		
♦ Collective buildings (Chapter VII)		3000	6000
♦ Single family houses (Chapter IV)		1700	3300
Road in tunnel (Chapter IV)	Total protection		
♦ 2 lanes tunnel		-	10.000-15.000
♦ 3 lanes tunnel		-	30.000-50.000

[1] Values obtained for single vehicles, 7.50 m from the axis of the rolling lane, 1.20 m from the ground

(o) 10 + 10 meters paved road
* Protection of top floors (see Figure VII.3)
** Especially at low frequencies
*** Considering the bi-annual cost of washing, and the eighth-year cost of recycling, actualized (without taking into account the inflation rate); the figure in parentheses indicates the pavement cost comparable with the other intervention methods that over the same period of time (15-16 years) have not required any form of intervention having costs similar to those of installation.
**** Difference between normal window and specialized "antinoise" window
***** Estimated for one house every 30 metres.

Soundproofing facades cost approximately $ 10 000 per living unit (French data). The range is from $ 6 000 for apartments in collective buildings, with an insulation of 30 dB(A), to $ 12 000 for a living unit in an isolated one or two-family building, for the same noise reduction level. Assuming that in an urban environment buildings have at least 3 floors, with 1 apartment per floor for every 10 m of road, the cost figure is $ 3 000 per linear m of road (naturally, for only one side; for two the cost would be $ 6 000).

All of these data are summed up in table VII.4, which is only indicative, and applicable on the basis of the conditions described in the text.

VII.4. REFERENCE

1. BAR, P and Y. DELANNE (1993). *Réduire le bruit pneumatique-chaussée*. Presse de l'Ecole Nationale des Ponts et Chaussées, Paris, France.

CHAPTER VIII

CONCLUSIONS, RECOMMENDATIONS AND RESEARCH NEEDS

VIII.1. NATIONAL APPROACHES

The preceding chapters were based on information extracted from an analysis of the national reports that were submitted. This analysis also provided material for assessing the basic noise strategies used in OECD countries. Although the national approaches are quite different, two main lines may be distinguished.

Countries such as Germany and France can be considered to have a historic awareness of the problem and for this reason have drawn up rules of conduct in the form of general laws and regulations which identify not only the limits to be achieved but also the technical and financial means necessary for obtaining the desired noise improvement. But, even in this group of countries, there is no uniformity in terms of procedure. Some countries, such as France, pursue improvements although remaining quite aware of the prevailing shortcomings of current technical methods. Research programmes have therefore been developed to resolve these issues based on a consolidated scientific approach.

Other countries with considerable experience of the problem, such as Switzerland, the Netherlands and Austria, consider that the actions taken for enforcing the rules on noise limits and their highly standardised measures on noise abatement are sufficient. They do not consider it necessary to elaborate major research programmes. Some also complain about the reluctance (common to other countries) in the adoption of non-subsidised protective measures or actions whose costs are borne by bodies other than road authorities. Such initiatives, entrusted to local authorities or private owners, account, in fact, only for a small portion of the total range of measures.

It is symptomatic for countries of the first type to exhibit the maximum effort to apply scientific rules in noise control, by defining measurement norms, forecasting models, new test features for anti-noise materials and structures with total protection. The countries of the second type, on the other hand, base their approach on standardised and very precise methodologies, which are applied without carrying out any numerical assessment of the results obtained. It should be noted however that the situations concerned are very much alike, with very few natural and climatic differences as to the physical environment. This kind of procedure allows indeed good cost forecasts, although the actual level of performance obtained is uncertain. The principal reasons are the lack of monitoring as well as the change of noise abatement properties of the facility during service, rather than lack of initial assessment of the validity of such structures.

Finally, there is another group of countries that are less advanced if one judges their position from the point of view whether or not they have adopted a general regulation or strategy based on established norms and earmarked budgets. These countries, such as Spain and Italy, proceed in a scientific way in choosing the technical treatments whereby the noise limits are pragmatically defined, based upon the experience of whoever first began to work in this field. The United States represents a separate case: there are not only fifty quite different situations, but also and principally because there is a different spirit as regards the regulation of community problems. While being highly advanced technically, the United States, in line with its liberal spirit, does not approach the problem through the use of legally fixed limits and plans. The Scandinavian countries constitute a slightly different case. Norway follows a "trial and error" method and has set up targets for noise reductions as guidelines for the planning act; different kinds of mitigation are used for noise reduction along new and existing roads to reduce "blackspots". Sweden clears up the consequences before action. Denmark is somewhat in between. For all these countries, however, the comparison of the different procedures developed and the assessment of their respective position have provided a useful basis for future national policies.

VIII.2. DIFFERENCE BETWEEN SYSTEMS FOR NEW AND EXISTING ROADS

All countries agree on the need to distinguish between noise control systems to use for new roads and those for existing roads.

VIII.2.1. New constructions

The overall goals are to reduce as much as possible the impacts of traffic noise and to optimise costs. In achieving these goals, the fundamental principles developed hereunder are generally held to be valid and the design development procedure should observe four principal phases:

a) A global approach to define the primary objectives in terms of the transport systems, regional area development plans, environment, financing and planning of scheme.

b) A preliminary study, so as to be able to intervene at the stage of infrastructure planning when the principal characteristics of the road are defined and the different variants are compared.

c) The actual route choice study including the determination of project characteristics.

d) The studies of details prior to the elaboration of the technical dossiers for the effective realisation of the works.

Each of these phases must include a section dedicated to noise pollution.

Global Approach

As regards noise impacts and like in the other sectors, this study covers the existing network and compares the different alternatives. In such important strategic choices, the noise criterion will rarely be determinant by itself but it can constitute an important element for regulatory policy.

The analysis will essentially be based upon traffic calculations and forecasts for the network in question. A summary map of the noise emitted by the infrastructure is prepared, taking into account the geometric layout, the traffic and, where possible, a qualitative evaluation of the surrounding built up areas and land uses.

When a decision is reached on the overall approach to be adopted for the area development plan and the comparisons of alternative schemes, it is not only necessary to evaluate the overall impacts of these different options, but also to ascertain the feasibility and cost of any protective structure needed. This is important because there is not always a direct relationship between the magnitude of the impacts to be contained and the magnitude of anti-noise structures. In fact, *a specific route alternative causing relatively limited nuisance -- even without any protective structure -- may reveal itself to be in absolute terms more difficult and more expensive to treat appropriately against noise than a basically more noise-polluting route alternative.*

The evaluation of the different impacts of the options considered must, however, not overlook the other important *effects affecting the existing network, such as increase or decrease of traffic flows, etc.*

Preliminary Studies

This is the stage at which all the regulatory provisions for minimising noise impacts of the proposed route, or its horizontal or vertical alignment (see Chapter IV), can be most easily incorporated into the design of the infrastructure. In general, these provisions bring about only limited cost increases insofar as they are incorporated in the design before implementation. If, however, noise impacts have not been taken into account at this stage, their subsequent handling will often mean that the project manager has to extend the anti-noise structures as expensive palliatives which will be more difficult to integrate in the site and more costly to maintain over time.

These studies require precise a priori definitions of the noise abatement objective[1] and the expected life of the structures -- 15 years for example --, which can vary according to the type of premises to protect (residential buildings, schools, hospitals, offices) or the nature of the primary noise generator. It is, furthermore, preferable to fix decision criteria for the noise measures selected. Priorities should be given to tackling pollution at its source in order to provide equal protection to the roadside areas -- whether public or private -- as well as to the application conditions of the protection measures for facades, etc. At this level, small scale acoustic studies are generally conducted, 1/5000 or 1/2000, using simple calculation methods, i.e. very often manual methods making use of nomograms or simplified formula. Only a few sites -- difficult topography, urban areas.. -- may call for larger scale studies or more sophisticated methods.

It is, however, indispensable to dispose of reliable traffic forecasts for the time horizons considered. The forecasts should not only focus on global traffic estimates but also include speeds, the proportion of heavy traffic (which has a big influence on the level of noise), night traffic and major seasonal variations, if any. By being essentially directed to problems of capacity and traffic flow, traffic forecast studies sometimes neglect some of these aspects. Current traffic forecast methods are still unsuitable for truck and night traffic predictions. Lacking accurate estimates for some of these parameters, it may be necessary to conduct specific studies on the existing network in order to obtain, through analogy, a sufficiently reliable basis for assessment.

1 For the desirable limits, see Chapter III and also Section VIII.3.2.

Furthermore, the study requires that along all the various routes foreseen, adequate knowledge of the topography exists, that sensitive areas have been identified and that buildings (year of construction, use) have been evaluated. Note also that the study entails consideration of a 400 to 600 metre wide band, which is far wider than the usual corridor for the analysis of the route itself. This aspect is very often neglected by the designers who concentrate their data collection on the immediate vicinity of the route and hence disregard the full effect of noise propagation.

Route Choice

Once the most suitable route has been chosen, a detailed study of its impact is carried out and the specific measures for noise protection selected. This includes:

- The choice between protection at source, facade treatment or mixed solutions;
- The nature and the dimensions of the protection work (hills, barriers, covers), height, length, facility, acoustic properties (in terms of transmission, absorption...);
- The principle of acoustic treatment of facades: the level of insulation needed, the type of treatment (double-glazing, closing up lodges, plastering facades), any constrains as regards ventilation and heating.

The objective of the study at this stage is to define the works and their geometric lay-out, to adopt all the conservation measures necessary, to estimate their cost and to make all the arrangements for the public enquiry.

The acoustic study and the design of the noise abatement measures should be realised with the help of detailed forecasting methods, usually making use of computers and sometimes 1/2000 or 1/500 scale models in particularly complex cases. The design route and/or other geometric design plans and studies provide the basic input data to have information about the surroundings, landuse, exposed openings, buildings facades, etc. as well as any precedents or particularities of the roadside premises that could affect forecasts and reactions to noise propagation.

Studies in Detail

Design studies of noise protection structures in the preliminary project phase must usually be taken up again at the final design stage in order to take account of:

- the definite geometric characteristics studied on a smaller scale;
- the reports and notices issued during the preliminary investigations and the arrangements for the public enquiry;
- the need to optimise the protection structures;
- any new requirements arising out of complementary and recent investigations with regard to the stability of structures, architectural and landscaping treatment, etc.

This is the occasion where the most reliable analysis means should be employed to take into account the complexity of the site. As mentioned above, sophisticated studies are carried out with the use of computers or by conducting tests on models.

Further detailed studies concern mainly architectural and landscaping treatments, the choice of materials and civil engineering studies concerning load-bearing structures, foundations, maintenance and safety. The studies should be conducted by a team of specialists in the various sectors, under the direction of the designer.

The Architecture of Noise Barriers[1]

In carrying out the architectural study it must be avoided that each noise barrier section be considered a particular operation to be studied separately, or even be sub-divided into sub-sections. On the other hand, it is not recommended to use too lively or ornamental colours since their aesthetics are often subject to debate.

The two faces of the barriers must be treated in a very different manner. On the side facing the traffic, the barrier is a road structure, and like the road, is a linear structure. It is not necessary, therefore, to be afraid of maintaining a horizontal linear structure, and there is no reason to wish to break it up unless there is the need to differentiate the carriageway itself. Furthermore, on a long journey, it is useful to have visual uniformity all along the route, and barriers can certainly play an important role in achieving this result.

On the external part of the barriers, however, the surface is visible to the outlying area and housings and thus the problem of enlivening it arises. This must be simply resolved by a method which sees the structure as an integral part of the landscaping process and not through occasional introduction of new elements. This is the typical road approach.

In Norway, on the contrary, while it is agreed that the two faces of the barriers must be treated in different manner, there is the position that the side facing the traffic should not be considered a barrier with a road structure. In this country, therefore, the different types of barriers have been grouped into three types: an area barrier, a garden screen and a local screen. These are different in architectural design depending on where they are situated and what type of form and function the road/street and the surroundings have. The area barrier which is used along motorways in suburban area screens an area and shall therefore reflect the area in its screening. It shall not be a part of the road but part of the area. In the urban area where garden or city screens are used, these screens will reflect the houses but also the street and will be part of the whole area.

In noise barrier tenders, it is strongly recommended that only the general architectural elements (materials, dimensions, etc.) be specified, and that each bidder be asked to employ the services of an architect in order to draw up a proposal adequate for the product. A design which is outlined in too precise details in the tender will often oblige suppliers to adapt their products in order to conform to these requirements, which will lead to a notable increase in costs. Another solution is to define designs for the barrier foundations and leave suppliers free to chose the panels that meet the precise acoustic requirements of the tender.

Public Involvement

The impact study which will determine the responsibilities of the transport/construction authority, may take place at different stages in the design phase. The best moment for public participation is after route choice, because the design is then sufficiently well defined to enable reliable impact evaluations and to propose adequate measures of protection. However, as the design is not definitively fixed it is possible to take into consideration the observations gathered in the course of the public enquiry whenever such a procedure is required by the laws of the country. In the decision process it is of course necessary to acquaint the population with the choices made in the phase preceding the public enquiry such as project development and route choice.

---

1 These recommendations are also valid for the measures on existing roads, as will be seen later.

As it is usually realised at the stage in which the route is chosen, the impact study can only provide a summary indication of the design and dimensions of the anti-noise structures. However, because the impact study forms part of the public enquiry process, it is necessary at this stage to state clearly what the chosen acoustic objectives and the principles of protection -- coverage, barrier, facade -- will be.

Computers and models offer new possibilities for communication to show the local community, in concrete form, how the principles and objectives of the preliminary studies will be translated into practice. The involvement of the local population may then become active and contribute to the decision-making process, for instance in deciding between a high barrier and facade treatment. This implication of the community is the best guarantee for the final success of the operation. Obviously, the information available must be honest and rigorous. It is especially dangerous to let it be believed that the barriers or berms will eliminate all the disturbance and completely hide the structure. The objective of this work should be understood as bringing the level of this disturbance down to a level held to be compatible with the existing residential development. It is also necessary to specify the types of noise involved and how they combine with other ambient noises.

When a barrier must be sited in the immediate vicinity of a built-up area, whether public (a street, a square, etc.) or private (garden), it is indispensable to involve the directly affected population in the architectural study. Alternatively, the risk would be run of seeing the design rejected, viewed unfavourably -- especially those whose property borders on the proposed new structure -- or even compromised. In conclusion, the procedure designed for new constructions also has the objective of showing that road noise abatement measures is not a pot pourri of different techniques. It is preferable to conceive an overall design in such a way that the best protection possible be ensured, while adjusting it to the landscape and the landuse.

VIII.2.2. Existing roads

As regards the protection of existing roads various opinions are held by Member countries:

- Some hold the view that existing roads should be left in present conditions;

- Others maintain that existing roads should receive noise protection if some form of modification -- e.g. enlargement -- is introduced but without taking action against the pre-existing noise levels.

- There is a third approach which advocates programmes for noise reduction along existing roads. Norway has applied this approach for about 15 years with a budget for noise mitigation to reduce noise along existing roads.

All countries agree that if protective measures are taken to control noise, the limits must necessarily be less stringent than those for new roads. This is clearly set out in the conclusions of chapter II where the various recommended levels for the two cases are reported.

The difficulty of making modifications to the conditions of existing roads -- which is the largest problem in most of the OECD countries -- concerns financing. Unlike new constructions or road widening covered by established budgets including the financing of noise abatement works, the funds available to the authorities responsible for existing roads are often insufficient, already even for ordinary road maintenance.

A recent study prepared by the European Community, however, foresees the introduction of additional noise limits for the elimination of *noise black spots*. This is a first level scenario and comprises measures for acoustic improvement. There are also second and third level scenarios: the former specifies limits which are almost identical to those recommended in this Report and the latter represents ideal limits to be pursued in the long term. This could mean that the obstacles mentioned above for the noise improvements of existing roads may be overcome. In E.U. countries, it might therefore be possible to take action on these roads by arranging appropriate financing to be entrusted to road administrations so that they can integrate anti-noise acoustic measures with ordinary maintenance actions which are normally implemented. An overall approach would then make it possible to enhance the roads by additional features, not just noise but also safety improvements for example. Chapter VI outlines this approach for some noise barriers and Chapter V for certain types of pavements. Measures of this sort are necessary to carry out to, first of all, eliminate certain noise black spots -- perhaps with the adoption of somewhat higher noise limits but lower than the current regulatory levels.

Before moving onto the discussion of the final recommendations, an important conclusion of this report needs to be highlighted. It concerns *maintenance* in its broad sense. Indeed, all the anti-noise measures which have been considered -- with the exception of those road structures set out in Chapter IV -- require regular inspection and maintenance in order to upkeep their initial acoustic properties. Therefore, emphasis should be on low cost methods for the measurement of acoustic conditions as well as the maintenance and restoration of those properties that may have been lost or compromised. The other conclusion concerns the need to apply, appropriately, the different types of mitigation depending on the function and layout of the road and its surroundings. Finally, it is recommended to set up a joint environmental team for dealing with these problems in the most comprehensive and economic way.

VIII.3. FINAL RECOMMENDATIONS

First of all, note that not all countries need to carry out all the recommendations developed hereunder and in this Report. Not only have many, as stated earlier, already put them into practice, but a large part of the recommendations stem in fact from the pioneering work of these countries. The Group has identified seven recommendations:

1. The need to institute general rules on acoustic levels, applicable to the whole national territory, is recognised. These should be well-defined levels, operational in terms of the evaluation methodology used and differentiated according to at least two areas, i.e. new roads and existing roads, the latter with "black spot" limits. Provisions for financing and implementing the measures are required at the same time.

2. "Supporting" rules for noise abatement programmes are important. "Noise respect" belts should be foreseen in town-planning and for all roads. In these areas, houses or buildings without noise protection cannot be foreseen. In other words, they must have an acoustic insulation and be constructed in such a way as to constitute barriers for other areas to be protected. In these belts the sound emission limits should be identical to those for the roads.

3. Noise pollution rules and limits should be periodically reviewed in order to verify their validity and technical and economic application. The progress and results of noise-abatement programmes should be monitored.

4. Fixed long monitoring facilities to evaluate the results of noise abatement, as indicated in the third recommendation, should be incorporated in the design of the road or in specific sections.

5. Also, and consequently, there is a need for the continuous improvement of noise-abatement forecasting models and noise measurement procedures both in regard to design and control. This can be achieved, at least partially, by exchanging information and adopting those models which are generally held to be the most appropriate, as set out in Chapter III and the attached bibliography.

6. Emphasis should be on the education of engineers and technical personnel to further study the development and maintenance of anti-noise systems. It is not, however, desirable to set up specialised courses. It is sufficient to include a specific subject "Road Noise Control", bringing together specialised knowledge in the framework of ordinary university and technical college courses.

7. Research on the entire range of treatments and measures to abate noise at source is recommended. Vehicle improvements and measures were outlined in the Report. In the next 10 years, measures should be taken on at least certain specific parts of vehicles which produce the more "difficult" noises from the point of view of present-day controls. These concern:

   - wheels,
   - exhaust pipes of goods vehicles, and
   - acceleration of goods vehicles.

   Improvements of these factors would most certainly make even the oldest forms of noise protection enormously more effective as well as making the abatement measures more economic in the long term.

VIII.4. RESEARCH IN PROGRESS AND RESEARCH NECESSARY

The various research areas concern the subjects mentioned in previous chapters. As a general note however, it can be observed that the research undertaken by the various countries seems to be shaped by prevailing national interests. For example, in the Nordic countries, the impact on society is the most important factor focusing on the social effects of noise. In the United States, the approach is pragmatic with manuals being prepared and disseminated while research in France and Italy is analytical with the aim on physical parameters, the evaluation of combined effects of different measures, etc.

This section of the Report does not present an exhaustive list of the research conducted in OECD countries, but an attempt to suggest a methodology based on the experience of the more advanced countries in certain sectors. Special attention should be paid to the coordination of this research, both to avoid duplication and make the results obtained by one country more readily available for use and/or further development by others. Examples of research efforts are cited to indicate what is most needed; they are grouped by type and not by country.

VIII.4.1. Current practices and noise limits

Most countries have set noise pollution limits. A few, including Denmark, Finland, and Norway, are conducting further impact studies. In Finland, current research is aimed at a better understanding of disturbance by noise as a part of disturbance by traffic generally. Norway has done research on the impact of traffic on the population's health and well-being. It was found, for example, that the percentage of people annoyed by traffic is greater in a large city than in a small town. The survey also showed that, until the traffic volume in an area as a whole becomes large, noise is the factor people find most annoying. Recent medical research indicates that people bothered by road traffic noise are more likely to become ill.

Against this background, the Danish research proposals are noteworthy:

- *Environmental improvements.* How can noise emission regulations be changed to improve the environment?

- *Adjusting noise limits to changing traffic trends.* Limit values are stated as equivalent continuous noise levels for 24 hours [LAeq(24)]. This was decided in the 1970s, when the equivalent level was thought to accurately reflect the nuisance to the population. But in the 1990s, the distribution of traffic over the 24 hours of the day may have changed, especially on motorways in Europe, possibly as a consequence of the new open market in Europe. There will be more heavy traffic at night, when people are asleep and more sensitive to noise, in particular noise peaks. Therefore a study should be undertaken to analyse the situation in view of developing new road noise guidelines.

- *Social costs.* Noise has a substantial impact on the social costs of road transport, which must be regarded as part of the true costs of transport. It would therefore be useful to undertake a study to assess the costs that road traffic noise entails for the society. In this context, it is perhaps noteworthy that a computer model to predict the total numbers of dwellings and people exposed to different levels of traffic noise has been developed in Norway (VSTOY).

Lastly, under this heading, reference can be made to the E.C. standard model, now used in many countries, to predict the air pollution of the given country. A similar model might be developed for noise pollution to allow comparison of the road noise situations in different countries.

VIII.4.2. Assessment and measurement

There is an increasing need for reliable models because actual measurements are time-consuming, and sometimes not possible. Research is being done in this area in many OECD countries. The following illustrates some example of this research. For example, in the United States, a Workshop was held in November 1991 to identify and prioritise environmental research needs in transport. Staff from State highway agencies, Federal Highway Administration (FHWA) and members of the highway traffic noise community developed highway traffic noise research problem statements on two major issues:

- *Development of a new highway traffic noise prediction procedure*
- *Field evaluation of traffic noise generation, propagation, and attenuation.*

The background to the first proposal -- traffic noise prediction -- relates to indications that the current programme overpredicts noise levels for certain situations. A correction of the overprediction, even by 1 dB(A), would reduce new barrier heights by approximately 0.6 metre, saving more than $ 10 million per year. The proposed research would involve three tasks:

- Review existing technology; evaluate the adequacy of that information; and, if not adequate, determine the extent of additional research needed.

- Develop/validate vehicle noise emission levels (including octave bands) and for heavy trucks on grades.

- Develop a new highway traffic noise prediction procedure and programme compatible with both the personal and mainframe computers. The programme will be designed with subsystems to allow information to be added now or in the future, when additional research is completed. A subsystem will be set up for each of the following areas:

  - Insertion loss - propagation over roadway pavement, ground, buildings, vegetation and atmospheric effects,
  - Multiple reflections,
  - Multiple diffractions,
  - Effects of barrier shapes,
  - Emission level source heights,
  - Octave band computations, and
  - Absorptive barriers.

It is planned to develop a complete and thorough user-friendly manual which will incorporate such items as minimum recommended barrier insertion loss, noisiest hour analysis, and barrier end-wrapping (The duration of the research foreseen is 24 months).

The problem statement on *Field evaluation of traffic noise generation, propagation and attenuation* notes that overprediction by the traffic noise prediction programme is caused in part by an uncertainty in the generation, propagation, and attenuation of traffic noise. Further field evaluation of several components of the programme is necessary to improve the accuracy of the prediction process. The proposed research will, through field testing, provide input to the noise prediction programme resulting in improvements in the accuracy and prediction of the generation, propagation, and attenuation of traffic noise. The following programme components, with a research duration of 48 months, will be field evaluated:

- Using the results of the review of existing technology from Problem 1 "Develop New Highway Traffic Noise Prediction Procedure", perform additional measurements required to evaluate propagation losses - propagation over roadway pavements, ground, buildings, vegetation, and atmospheric effects.

- Measuring the acoustical performance of a variety of staged barriers options along suitable roadway location. As a minimum, the design options to be measured will include parallel reflective barriers and absorptive barriers with a minimum Noise Reduction Coefficient of 0.8.

- Better predicting diffraction over noise barriers; vehicle component source heights and relative emission levels as a function of speed will be identified so the diffracted noise of each component can be calculated by the new highway traffic noise prediction programme.

- Uncertainty exists about the longevity of the acoustical benefits of open graded pavements. A long-term measurement study will be undertaken to evaluate the degradation in pavement acoustical performance over time until the benefits are 1 dB(A) or less.

The Federal Highway Administration has a research underway to address the first two problem statements as listed above. A review of existing national and international technical knowledge related to highway traffic noise has been completed. Currently, a new highway traffic noise prediction model with implementing microcomputer software is being developed to incorporate state-of-the-art advances in vehicle noise emission levels, as well as in traffic noise prediction, barrier analysis and design, and computer technology. The new model and software should be available by the end of 1995.

In Switzerland, the federal laboratory EMPA currently conducts traffic noise research aiming at a *better computing model*. Specifically, better predictions are needed in towns. Also, the current model does not take reflections into account. Another research field relates to the *sound absorption coefficient* required for a wall or other acoustic barrier and its important influence on actual costs. However there is no reliable way of predicting the on-site absorption coefficient and EMPA has therefore developed an impulse echo method.

In France (Laboratoire Central des Ponts et Chaussées), the research programme on outdoor noise emission and propagation prediction methods includes:

- *Qualification of emissions for various kinds of pavements.* This covers the identification of the acoustic power of various types of vehicle on various classes of pavement at various speeds. These values will make it possible to classify pavements in acoustic terms as a function of speed. The Franco-German method will be used for this experimental study.

- *Influence of meteorological effects on tire/road noise propagation.* In addition to major research on the effects of atmospheric conditions on outdoor noise propagation, there will be special emphasis on the particular effects of line sources. Is there a great difference between a line source and a point source.

- *Influence of topography on traffic noise propagation.*

- *Comparison of various prediction models.* Validation of different countries' prediction models on three or four typical cases.

- *Calculation of noise-barrier/low-noise-pavement interaction.* Calculation of combined means of protection.

VIII.4.3. Anti-noise design and layout

Environmental impact assessments are a routine part of new road planning. An international study of the various ways of noise valuation for this purpose would be benefical and point to improvements of various national methods. This could also lead to a more strategic way to approach the subject.

Another problem concerns the effects that different kinds of noise reduction have on people. If a noise barrier is erected that has a negative effect on people, even if it provides a substantial noise reduction, it is necessary to ask whether this is the right solution. Perhaps other ways or means of mitigation may give better effects on people's attitudes to roads and road traffic, even if the noise reduction is not so effective. Some countries have therefore undertaken research on this subject, using various methodologies. If countries had a similar approach, it could lead to a better understanding of the total road and traffic effect on people and of the kind of mitigation that will be most effective including the consideration of social effect.

The proposed research on *People annoyance of traffic noise* will therefore investigate the effects of different mitigations in terms of fewer people that will be annoyed by noise. Research so far has been conducted in different countries to try to find out how many people were annoyed by traffic noise and to what level. However these studies are not comparable with each other. Consequently, the research tasks (duration of 24 months) will be:

- Define the same basic approach to the question of people annoyance
- Compare and analyse the research undertaken
- Find the effect of people annoyance before and after a measure is implemented
- Study annoyance impacts of different kinds of mitigation.

One measure that has shown a good impact on noise reduction is the development and use of *less noisy tyres*. Tyre manufacturers are doing considerable research on noise reduction. Today, automobile industry's demand for quiet tyres is very strong. At higher speeds (> 50 km/h) the tyre-road noise is higher than the emission noise of the vehicles. The aim is therefore to reduce the emissions generated at the contact surface of tyre and road.

Many research studies have been undertaken in the field of low noise pavement, but there are other possibilities. Research, undertaken in Sweden by Sandberg, showed that another *concept of the wheel* may give higher reduction of noise emissions. Reductions by up to 10 dB(A) should be possible and this would represent a reduction by 50 per cent of the noise emission of vehicles. If this is realised further noise emissions on the other sources of the vehicle are useful. A 10 dB(A) reduction is considerable if we compare this figure to the effects of other anti-noise measures:

- Normal porous asphalt: 2-4 dB(A)
- Double layer porous asphalt: 5-7 dB(A)
- Well designed noise barrier: 7 dB(A)
- Well designed double glazing: 10 dB(A).

At this moment, it is known that the driving properties of plastic wheels are different from normal wheels. Further development of this concept is recommended to achieve an extra 10 dB(A) noise reduction in the future.

VIII.4.4. Low noise road surfacings

Phonoabsorbent draining pavements are the most popular way to control noise. But several countries are reluctant to generalise their use, because of lack of experience and research. There is relative scarcity of information on factors such as durability, long-term effectiveness, winter safety, and declogging effectiveness of the draining layers.

In Italy, eight million square meters of phonoabsorbent pavements, equivalent to about 800 km of road, have been tested on the network operated by the Autostrade company. All of the work done clearly shows that draining pavements are effective only in reducing medium or high frequencies of vehicle noise emissions. It is important to discover "active" methods capable of damping low-frequency components. Resonators are generally used to absorb low-frequency sound. The possibility was accordingly examined of incorporating elements in the road base to make the pavement phonoabsorbent even at low frequencies. Prototypes of such pavements are undergoing laboratory tests.

VIII.4.5. Noise barriers

Most research on barriers seems to be done by private companies. In Switzerland and in other countries, governmental laboratories, such as EMPA (in Switzerland), conduct tests to certify the quality and specifications of products.

In Italy, the Autostrade company has tested various types of barriers (biowall, concrete or wood absorptive transparent wall), but more original is research on vegetation barriers and New Jersey low frequency absorptive barriers. First prototypes of New Jersey phonoabsorbent barriers with a selective configuration to absorb low frequencies have given excellent performance between 100 and 500 Hz (1992).

A series of noise measurement surveys was carried out in 1987-88 and the sound attenuation properties of various plant species measured. The best barrier for elevated roadways, for example, was found to be a thick evergreen hedge combined with several rows of trees. In 1990 and 1991 it was shown that extremely compact vegetation barriers roughly 9 to 10 meters thick (and never less than 6-7 m) can yield 3-4 dB(A) of noise abatement; but the effectiveness of such a barrier falls off with increasing distance, completely disappearing 80 to 100 m from the road. The tests showed that a vegetation barrier filters out proportionally more high-frequency noise, making traffic noise less disturbing for a given Leq. More particularly, the laboratory research under way is aimed at quantifying, in a series of sound chamber tests, the phonoabsorbent characteristics of certain species used in noise barriers.

A second, related, series of tests involves specific measurements on leaf samples from the same plants, using standing wave methods. The tests were performed in sound chambers in Italy. A comparison of the measurements on the plants with and without foliage shows that the branches and trunks have virtually no effect, and the difference between the two measurements is, and can only be, attributed to the leaves. The influence of plant density was also checked. A clear direct relation between plant density and acoustic absorption is found. The obvious drawback of vegetation barriers is that they take a long time to reach full effectiveness. Because of this, we do not yet have complete acoustic data on vegetation barriers designed and created solely for noise abatement. .

U.S. priorities in this field are the following:

- *Highway Traffic Noise Research Technology Transfer Via Handbooks and Training Aids.*
- *Highway Noise - Synthesis and Development of Standard Methods and Guidance Packages.*

As to the first topic, the US expert state that over the years, much excellent highway traffic noise research had been completed by many different government and private researchers. The results of this research have however not been consistently shared. Therefore, much duplication of effort has resulted in a large cost impact on some government agencies. In those cases where the information is shared, it frequently is in the form of a final report that has limited distribution and requires significant interpretation of data. If these data were converted to useful handbooks and/or training aids, the vast quantity of highway traffic noise experience could be better shared. This will result in improved noise barrier designs and better trained personnel responsible for the development, construction and maintenance of noise barriers. It will also result in real time and cost savings by eliminating duplication of effort.

The project (of a duration of 30 months) aims at improving the transfer of technology among the user community. The task is to *update the FHWA Noise Barrier Design Handbook*, AASHTO's (American Association of State Highway and Transportation Officials) *Noise Barrier Design Guide*, and the recently completed NCHRP (National Cooperative Highway Research Program) *synthesis of noise barrier practice*. This update will be a how-to document that is a compendium of easily understood design options that a structural designer with little or no design experience can turn to. It will provide design details on a series of frequently used barrier types (wood, metal, concrete) as well as innovative materials. It will guide the designer through the pitfalls that experienced designers would anticipate, such as cost effective design concepts for handling drainage considerations, dealing with fire protection needs, and providing maintenance access without diminishing the noise barrier's acoustical performance. A videotape version of the updated FHWA Noise Barrier Design Handbook will be developed as well as a series of training aids that can be distributed to the user community in the areas of noise barrier construction and maintenance.

Since opportunities for construction noise mitigation are often overlooked, the second project (of a duration of 36 months) will synthesise common practice, perform a rigorous and thorough evaluation of this synthesis, and develop methods/guidance documents, as follows:

- Refine and tighten the *FHWA's "Sound Procedure for Measuring Highway Noise"*.
- Develop and field test a *survey instrument* to measure attitudes, both positive and negative, about highway noise barriers. Of most interest are the attitudes of neighbours who directly abut a noise barrier -- i.e., those neighbours affected most by the barrier's visual appearance and benefitted most by its noise reduction. Use the professional knowledge and skills of survey sociologists to measure the influence of all independent variables, to avoid survey bias, and to ensure a high degree of statistical reliability.
- Develop a flexible, generic presentation package for *public involvement*.
- Synthetise, evaluate current *highway construction noise abatement* techniques, and publish a series of prototype construction noise specifications.

In Denmark, somewhat similar survey research is advocated, since knowledge of what people living near roads really think about noise barriers is spotty, as is knowledge of drivers' opinions of noise barriers. There may be a correlation between noise barriers, drivers' behaviour, and traffic safety. Further research on these points might provide a stronger background for the development of noise barrier design guidelines.

VIII.4.6. Future research ideas

The closing part of the recommendations in section VIII.3 as well as the preceding discussions show that the areas open to investigate noise control are still very wide. In this context, it is not unreasonable to expect that electronic and microprocessor developments could lead to economically valid solutions for road noise abatement.

An emblematic example of innovative ideas is the use of "counter-noise" technology in which a specific sound frequency spectrum is generated in order to interfere with fastidious noise emissions. Research in this sector is still in its infancy and addresses more the phenomenon of vibration than acoustic phenomena as such, focusing on closed rather than open environments. However, it could now be possible to reach substantial improvements in the control of goods vehicles exhausts.

LIST OF PARTICIPANTS

Chairman: Mr Gabriele CAMOMILLA

AUSTRALIA	Mr R. MATTHEWS
AUSTRIA	Mr F. ZOTTER
FINLAND	Mr A. JANSSON
FRANCE	Mr Y. DELANNE Mr M. BERENGIER
ITALY	Mr G. CAMOMILLA Mr M. LUMINARI Mr S. GERVASIO Mrs P. PICHETTI
JAPAN	Mr M. ISHIDA Mr K. MORI Mr T. NAKAMURA Mr T. SAITO
NETHERLANDS	Mr R.M. DURA Mr C. PADMOS Mr H.D. VAN BOHEMEN
NORWAY	Mrs A. MARSTEIN
SPAIN	Mr J. TRIGUEROS RODRIGO Mr F. RUZA TARRIO
SWITZERLAND	Mr L. FROIDEVAUX
UNITED KINGDOM	Mr P. NELSON
UNITED STATES	Mr R. ARMSTRONG
OECD	Mr B. HORN Mr C. MORIN Ms V. FEYPELL

Rapporteurs of the Chapters of the Report were Mrs Marstein and Messrs R. Armstrong, G. Camomilla, Y. Delanne, L. Froidevaux, S. Gervasio, M. Luminari, J. Trigueros. The final report was co-ordinated by the OECD Secretariat.

MAIN SALES OUTLETS OF OECD PUBLICATIONS
PRINCIPAUX POINTS DE VENTE DES PUBLICATIONS DE L'OCDE

ARGENTINA – ARGENTINE
Carlos Hirsch S.R.L.
Galería Güemes, Florida 165, 4° Piso
1333 Buenos Aires Tel. (1) 331.1787 y 331.2391
Telefax: (1) 331.1787

AUSTRALIA – AUSTRALIE
D.A. Information Services
648 Whitehorse Road, P.O.B 163
Mitcham, Victoria 3132 Tel. (03) 873.4411
Telefax: (03) 873.5679

AUSTRIA – AUTRICHE
Gerold & Co.
Graben 31
Wien I Tel. (0222) 533.50.14
Telefax: (0222) 512.47.31.29

BELGIUM – BELGIQUE
Jean De Lannoy
Avenue du Roi 202 Koningslaan
B-1060 Bruxelles Tel. (02) 538.51.69/538.08.41
Telefax: (02) 538.08.41

CANADA
Renouf Publishing Company Ltd.
1294 Algoma Road
Ottawa, ON K1B 3W8 Tel. (613) 741.4333
Telefax: (613) 741.5439
Stores:
61 Sparks Street
Ottawa, ON K1P 5R1 Tel. (613) 238.8985
211 Yonge Street
Toronto, ON M5B 1M4 Tel. (416) 363.3171
Telefax: (416)363.59.63

Les Éditions La Liberté Inc.
3020 Chemin Sainte-Foy
Sainte-Foy, PQ G1X 3V6 Tel. (418) 658.3763
Telefax: (418) 658.3763

Federal Publications Inc.
165 University Avenue, Suite 701
Toronto, ON M5H 3B8 Tel. (416) 860.1611
Telefax: (416) 860.1608

Les Publications Fédérales
1185 Université
Montréal, QC H3B 3A7 Tel. (514) 954.1633
Telefax: (514) 954.1635

CHINA – CHINE
China National Publications Import
Export Corporation (CNPIEC)
16 Gongti E. Road, Chaoyang District
P.O. Box 88 or 50
Beijing 100704 PR Tel. (01) 506.6688
Telefax: (01) 506.3101

CHINESE TAIPEI – TAIPEI CHINOIS
Good Faith Worldwide Int'l. Co. Ltd.
9th Floor, No. 118, Sec. 2
Chung Hsiao E. Road
Taipei Tel. (02) 391.7396/391.7397
Telefax: (02) 394.9176

CZECH REPUBLIC – RÉPUBLIQUE TCHÈQUE
Artia Pegas Press Ltd.
Narodni Trida 25
POB 825
111 21 Praha 1 Tel. 26.65.68
Telefax: 26.20.81

DENMARK – DANEMARK
Munksgaard Book and Subscription Service
35, Nørre Søgade, P.O. Box 2148
DK-1016 København K Tel. (33) 12.85.70
Telefax: (33) 12.93.87

EGYPT – ÉGYPTE
Middle East Observer
41 Sherif Street
Cairo Tel. 392.6919
Telefax: 360-6804

FINLAND – FINLANDE
Akateeminen Kirjakauppa
Keskuskatu 1, P.O. Box 128
00100 Helsinki
Subscription Services/Agence d'abonnements :
P.O. Box 23
00371 Helsinki Tel. (358 0) 121 4416
Telefax: (358 0) 121.4450

FRANCE
OECD/OCDE
Mail Orders/Commandes par correspondance:
2, rue André-Pascal
75775 Paris Cedex 16 Tel. (33-1) 45.24.82.00
Telefax: (33-1) 49.10.42.76
Telex: 640048 OCDE
Internet: Compte.PUBSINQ @ oecd.org
Orders via Minitel, France only/
Commandes par Minitel, France exclusivement :
36 15 OCDE
OECD Bookshop/Librairie de l'OCDE :
33, rue Octave-Feuillet
75016 Paris Tel. (33-1) 45.24.81.81
(33-1) 45.24.81.67

Documentation Française
29, quai Voltaire
75007 Paris Tel. 40.15.70.00
Gibert Jeune (Droit-Économie)
6, place Saint-Michel
75006 Paris Tel. 43.25.91.19
Librairie du Commerce International
10, avenue d'Iéna
75016 Paris Tel. 40.73.34.60
Librairie Dunod
Université Paris-Dauphine
Place du Maréchal de Lattre de Tassigny
75016 Paris Tel. (1) 44.05.40.13
Librairie Lavoisier
11, rue Lavoisier
75008 Paris Tel. 42.65.39.95
Librairie L.G.D.J. - Montchrestien
20, rue Soufflot
75005 Paris Tel. 46.33.89.85
Librairie des Sciences Politiques
30, rue Saint-Guillaume
75007 Paris Tel. 45.48.36.02
P.U.F.
49, boulevard Saint-Michel
75005 Paris Tel. 43.25.83.40
Librairie de l'Université
12a, rue Nazareth
13100 Aix-en-Provence Tel. (16) 42.26.18.08
Documentation Française
165, rue Garibaldi
69003 Lyon Tel. (16) 78.63.32.23
Librairie Decitre
29, place Bellecour
69002 Lyon Tel. (16) 72.40.54.54
Librairie Sauramps
Le Triangle
34967 Montpellier Cedex 2 Tel. (16) 67.58.85.15
Tekefax: (16) 67.58.27.36

GERMANY – ALLEMAGNE
OECD Publications and Information Centre
August-Bebel-Allee 6
D-53175 Bonn Tel. (0228) 959.120
Telefax: (0228) 959.12.17

GREECE – GRÈCE
Librairie Kauffmann
Mavrokordatou 9
106 78 Athens Tel. (01) 32.55.321
Telefax: (01) 32.30.320

HONG-KONG
Swindon Book Co. Ltd.
Astoria Bldg. 3F
34 Ashley Road, Tsimshatsui
Kowloon, Hong Kong Tel. 2376.2062
Telefax: 2376.0685

HUNGARY – HONGRIE
Euro Info Service
Margitsziget, Európa Ház
1138 Budapest Tel. (1) 111.62.16
Telefax: (1) 111.60.61

ICELAND – ISLANDE
Mál Mog Menning
Laugavegi 18, Pósthólf 392
121 Reykjavik Tel. (1) 552.4240
Telefax: (1) 562.3523

INDIA – INDE
Oxford Book and Stationery Co.
Scindia House
New Delhi 110001 Tel. (11) 331.5896/5308
Telefax: (11) 332.5993
17 Park Street
Calcutta 700016 Tel. 240832

INDONESIA – INDONÉSIE
Pdii-Lipi
P.O. Box 4298
Jakarta 12042 Tel. (21) 573.34.67
Telefax: (21) 573.34.67

IRELAND – IRLANDE
Government Supplies Agency
Publications Section
4/5 Harcourt Road
Dublin 2 Tel. 661.31.11
Telefax: 475.27.60

ISRAEL
Praedicta
5 Shatner Street
P.O. Box 34030
Jerusalem 91430 Tel. (2) 52.84.90/1/2
Telefax: (2) 52.84.93

R.O.Y. International
P.O. Box 13056
Tel Aviv 61130 Tel. (3) 546 1423
Telefax: (3) 546 1442

Palestinian Authority/Middle East:
INDEX Information Services
P.O.B. 19502
Jerusalem Tel. (2) 27.12.19
Telefax: (2) 27.16.34

ITALY – ITALIE
Libreria Commissionaria Sansoni
Via Duca di Calabria 1/1
50125 Firenze Tel. (055) 64.54.15
Telefax: (055) 64.12.57
Via Bartolini 29
20155 Milano Tel. (02) 36.50.83
Editrice e Libreria Herder
Piazza Montecitorio 120
00186 Roma Tel. 679.46.28
Telefax: 678.47.51

Libreria Hoepli
Via Hoepli 5
20121 Milano Tel. (02) 86.54.46
Telefax: (02) 805.28.86

Libreria Scientifica
Dott. Lucio de Biasio 'Aeiou'
Via Coronelli, 6
20146 Milano Tel. (02) 48.95.45.52
Telefax: (02) 48.95.45.48

JAPAN – JAPON
OECD Publications and Information Centre
Landic Akasaka Building
2-3-4 Akasaka, Minato-ku
Tokyo 107 Tel. (81.3) 3586.2016
Telefax: (81.3) 3584.7929

KOREA – CORÉE
Kyobo Book Centre Co. Ltd.
P.O. Box 1658, Kwang Hwa Moon
Seoul Tel. 730.78.91
Telefax: 735.00.30

MALAYSIA – MALAISIE
University of Malaya Bookshop
University of Malaya
P.O. Box 1127, Jalan Pantai Baru
59700 Kuala Lumpur
Malaysia Tel. 756.5000/756.5425
Telefax: 756.3246

MEXICO – MEXIQUE
Revistas y Periodicos Internacionales S.A. de C.V.
Florencia 57 - 1004
Mexico, D.F. 06600 Tel. 207.81.00
Telefax: 208.39.79

NETHERLANDS – PAYS-BAS
SDU Uitgeverij Plantijnstraat
Externe Fondsen
Postbus 20014
2500 EA's-Gravenhage Tel. (070) 37.89.880
Voor bestellingen: Telefax: (070) 34.75.778

NEW ZEALAND
NOUVELLE-ZÉLANDE
GPLegislation Services
P.O. Box 12418
Thorndon, Wellington Tel. (04) 496.5655
Telefax: (04) 496.5698

NORWAY – NORVÈGE
Narvesen Info Center – NIC
Bertrand Narvesens vei 2
P.O. Box 6125 Etterstad
0602 Oslo 6 Tel. (022) 57.33.00
Telefax: (022) 68.19.01

PAKISTAN
Mirza Book Agency
65 Shahrah Quaid-E-Azam
Lahore 54000 Tel. (42) 353.601
Telefax: (42) 231.730

PHILIPPINE – PHILIPPINES
International Book Center
5th Floor, Filipinas Life Bldg.
Ayala Avenue
Metro Manila Tel. 81.96.76
Telex 23312 RHP PH

PORTUGAL
Livraria Portugal
Rua do Carmo 70-74
Apart. 2681
1200 Lisboa Tel. (01) 347.49.82/5
Telefax: (01) 347.02.64

SINGAPORE – SINGAPOUR
Gower Asia Pacific Pte Ltd.
Golden Wheel Building
41, Kallang Pudding Road, No. 04-03
Singapore 1334 Tel. 741.5166
Telefax: 742.9356

SPAIN – ESPAGNE
Mundi-Prensa Libros S.A.
Castelló 37, Apartado 1223
Madrid 28001 Tel. (91) 431.33.99
Telefax: (91) 575.39.98

Libreria Internacional AEDOS
Consejo de Ciento 391
08009 – Barcelona Tel. (93) 488.30.09
Telefax: (93) 487.76.59

Llibreria de la Generalitat
Palau Moja
Rambla dels Estudis, 118
08002 – Barcelona
(Subscripcions) Tel. (93) 318.80.12
(Publicacions) Tel. (93) 302.67.23
Telefax: (93) 412.18.54

SRI LANKA
Centre for Policy Research
c/o Colombo Agencies Ltd.
No. 300-304, Galle Road
Colombo 3 Tel. (1) 574240, 573551-2
Telefax: (1) 575394, 510711

SWEDEN – SUÈDE
Fritzes Customer Service
S–106 47 Stockholm Tel. (08) 690.90.90
Telefax: (08) 20.50.21

Subscription Agency/Agence d'abonnements :
Wennergren-Williams Info AB
P.O. Box 1305
171 25 Solna Tel. (08) 705.97.50
Telefax: (08) 27.00.71

SWITZERLAND – SUISSE
Maditec S.A. (Books and Periodicals - Livres et périodiques)
Chemin des Palettes 4
Case postale 266
1020 Renens VD 1 Tel. (021) 635.08.65
Telefax: (021) 635.07.80

Librairie Payot S.A.
4, place Pépinet
CP 3212
1002 Lausanne Tel. (021) 341.33.47
Telefax: (021) 341.33.45

Librairie Unilivres
6, rue de Candolle
1205 Genève Tel. (022) 320.26.23
Telefax: (022) 329.73.18

Subscription Agency/Agence d'abonnements :
Dynapresse Marketing S.A.
38 avenue Vibert
1227 Carouge Tel. (022) 308.07.89
Telefax: (022) 308.07.99

See also – Voir aussi :
OECD Publications and Information Centre
August-Bebel-Allee 6
D-53175 Bonn (Germany) Tel. (0228) 959.120
Telefax: (0228) 959.12.17

THAILAND – THAÏLANDE
Suksit Siam Co. Ltd.
113, 115 Fuang Nakhon Rd.
Opp. Wat Rajbopith
Bangkok 10200 Tel. (662) 225.9531/2
Telefax: (662) 222.5188

TURKEY – TURQUIE
Kültür Yayinlari Is-Türk Ltd. Sti.
Atatürk Bulvari No. 191/Kat 13
Kavaklidere/Ankara Tel. 428.11.40 Ext. 2458
Dolmabahce Cad. No. 29
Besiktas/Istanbul Tel. (312) 260 7188
Telex: (312) 418 29 46

UNITED KINGDOM – ROYAUME-UNI
HMSO
Gen. enquiries Tel. (171) 873 8496
Postal orders only:
P.O. Box 276, London SW8 5DT
Personal Callers HMSO Bookshop
49 High Holborn, London WC1V 6HB
Telefax: (171) 873 8416
Branches at: Belfast, Birmingham, Bristol, Edinburgh, Manchester

UNITED STATES – ÉTATS-UNIS
OECD Publications and Information Center
2001 L Street N.W., Suite 650
Washington, D.C. 20036-4910 Tel. (202) 785.6323
Telefax: (202) 785.0350

VENEZUELA
Libreria del Este
Avda F. Miranda 52, Aptdo. 60337
Edificio Galipán
Caracas 106 Tel. 951.1705/951.2307/951.1297
Telegram: Libreste Caracas

Subscription to OECD periodicals may also be placed through main subscription agencies.

Les abonnements aux publications périodiques de l'OCDE peuvent être souscrits auprès des principales agences d'abonnement.

Orders and inquiries from countries where Distributors have not yet been appointed should be sent to: OECD Publications Service, 2 rue André-Pascal, 75775 Paris Cedex 16, France.

Les commandes provenant de pays où l'OCDE n'a pas encore désigné de distributeur peuvent être adressées à : OCDE, Service des Publications, 2, rue André-Pascal, 75775 Paris Cedex 16, France.

7-1995

OECD PUBLICATIONS, 2 rue André-Pascal, 75775 PARIS CEDEX 16
PRINTED IN FRANCE
(77 95 05 1) ISBN 92-64-14578-8 - No. 48199 1995